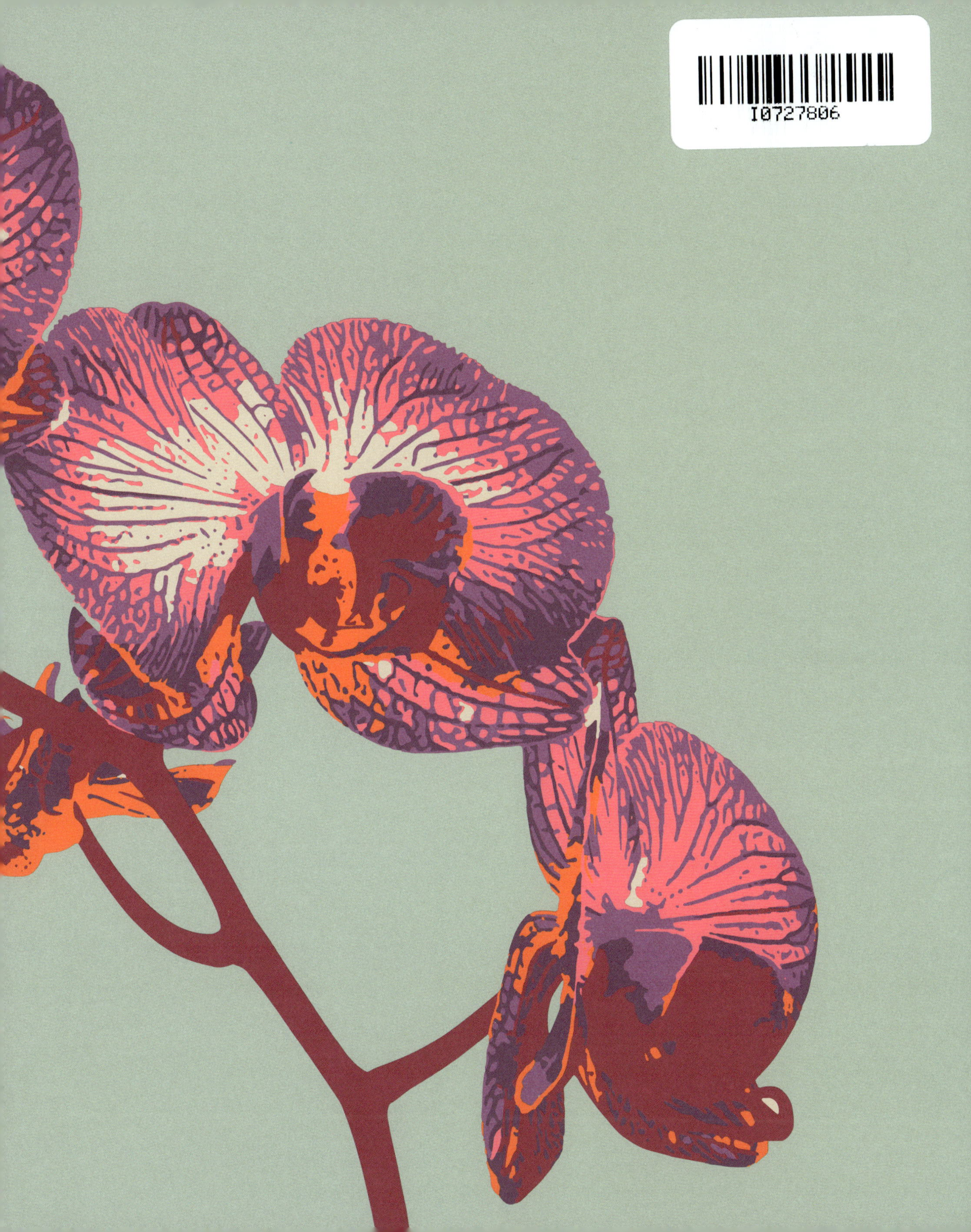

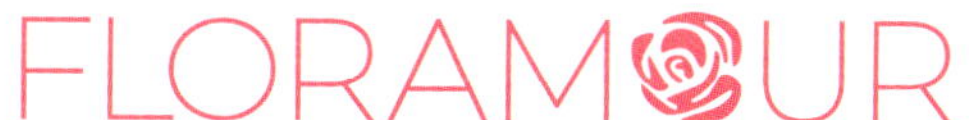

ANJA KLAFFENBACH

ORCHIDS
ORCHIDEEN

teNeues

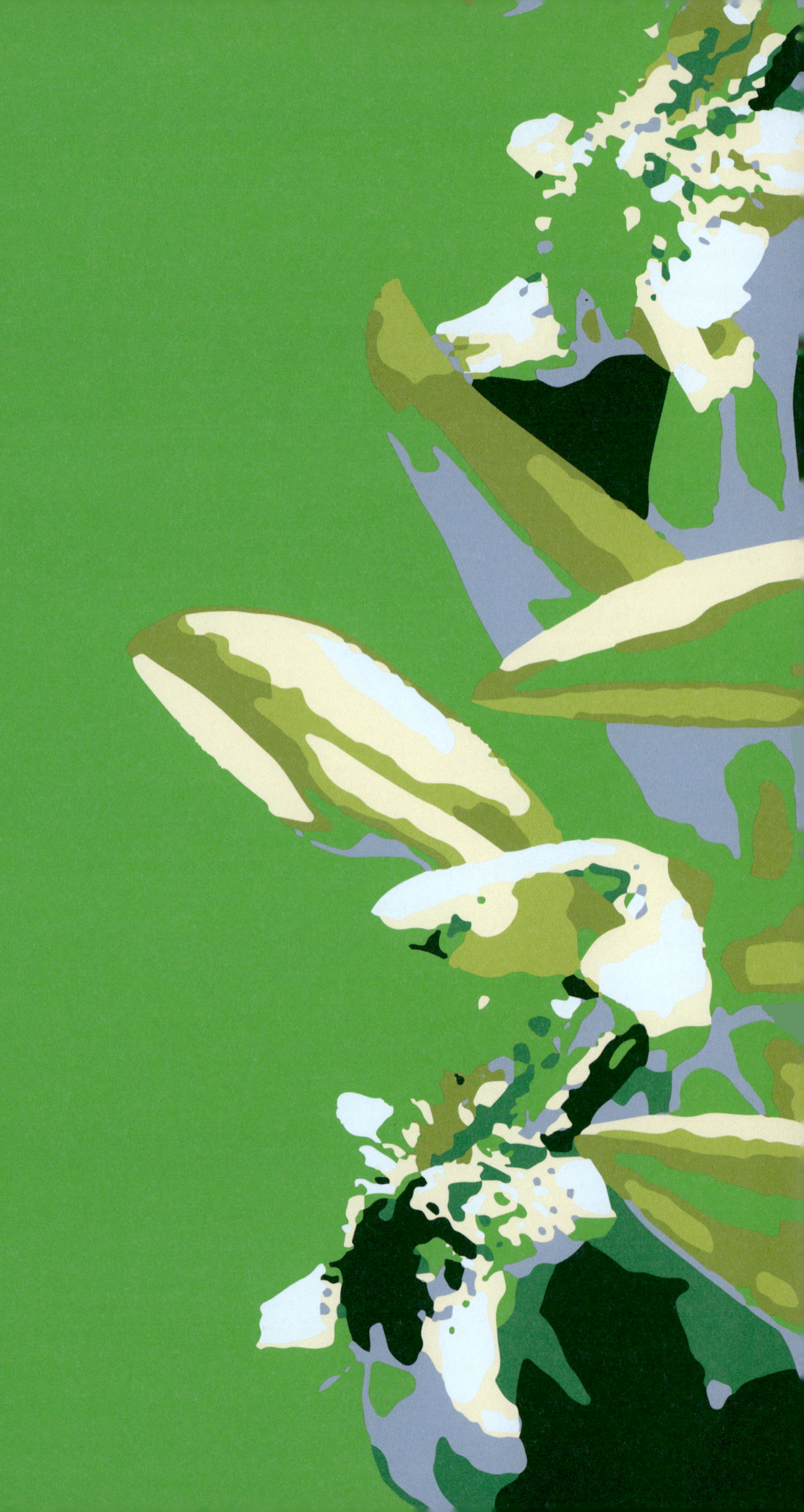

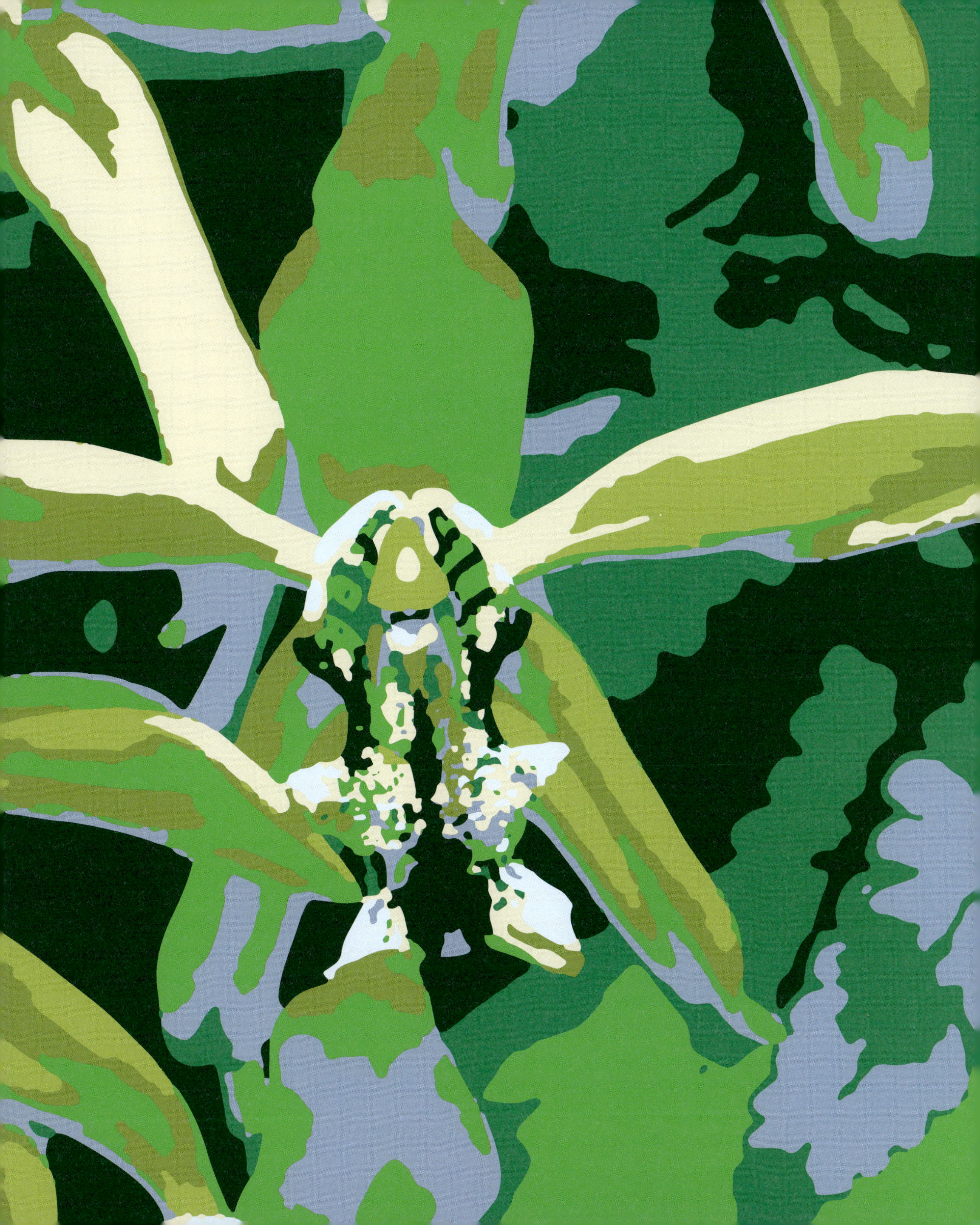

CONTENT
INHALT

FASCINATING BOTANY
FASZINIERENDE BOTANIK

MASTERPIECES OF NATURE
MEISTERWERKE DER NATUR

CULTURE, COSMETICS AND CUISINE
KULTUR, KOSMETIK & KULINARIK

ORCHIDS THROUGHOUT HISTORY
GESCHICHTE UND GESCHICHTEN

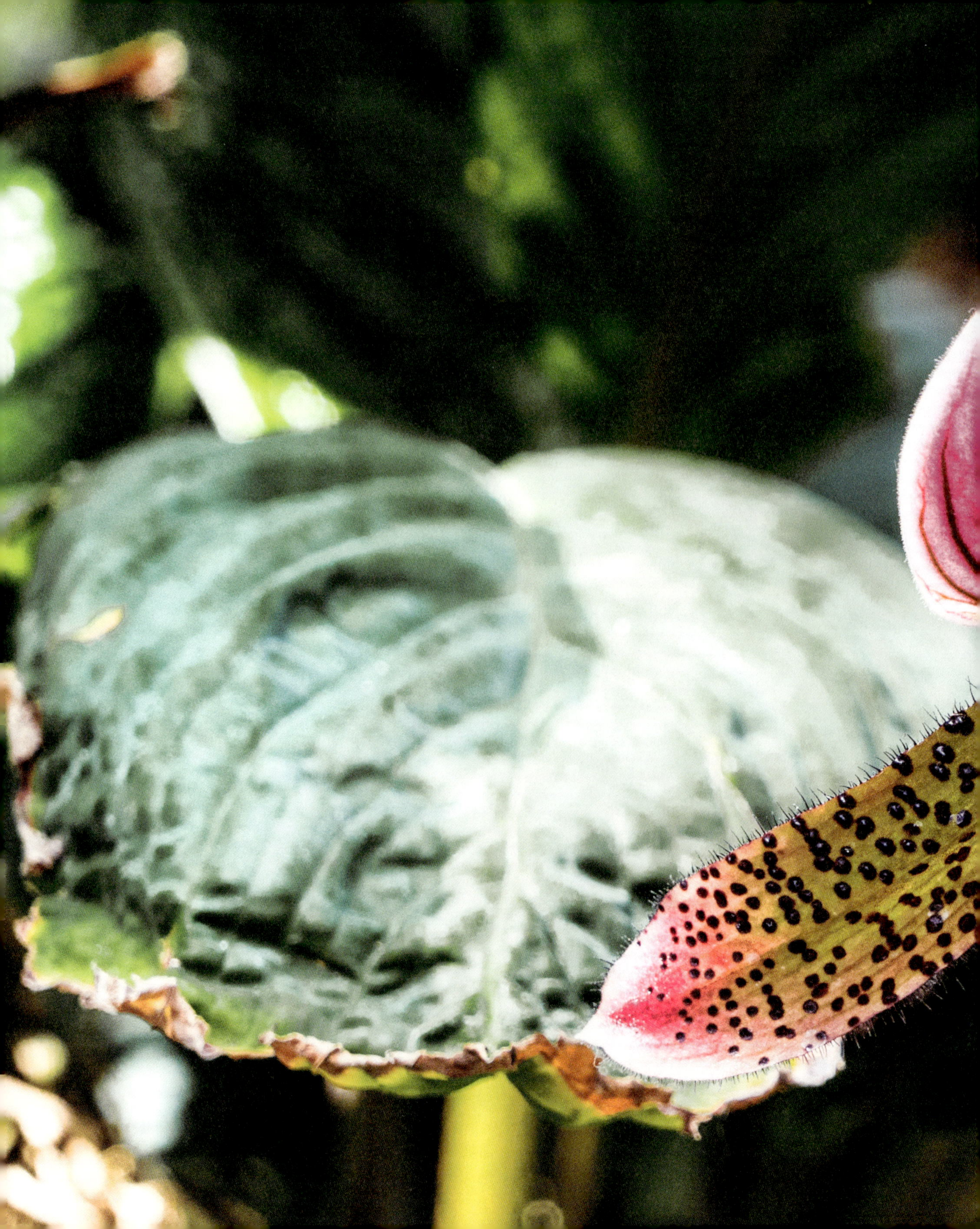

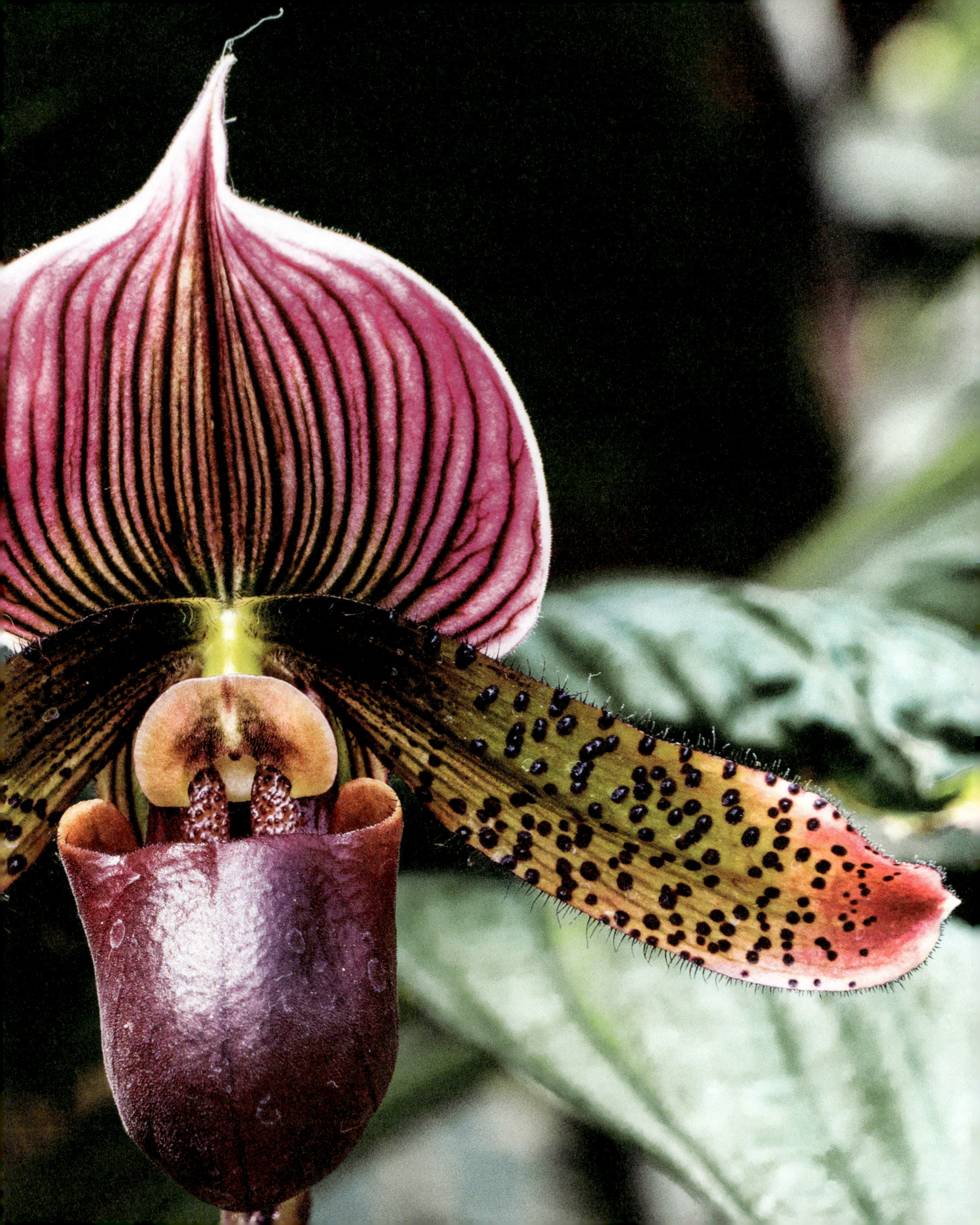

TREASURES TO DISCOVER AND PRESERVE

As a child, I never would have thought it possible that there are around 30,000 different species of orchids in the world. In fact, for many years, the only orchid I would encounter was the same old *Phalaenopsis* that seemed to be in every family and friend's living room. It wasn't until I started visiting various botanical gardens, particularly the stunning greenhouses at Kew Gardens, that I began to appreciate the many genera and species of tropical orchids waiting to be discovered. This marked the beginning of a growing fascination with these beautiful plants.

And it was not until I got involved with the conservation efforts of the organization Arbeitskreise Heimische Orchideen (Native orchids working groups) that I learned about the enchanting orchid species right in my own backyard. Each year, the organization highlights particularly endangered species by selecting an "Orchid of the Year." May this book help draw attention to these valuable natural treasures and their need for protection.

In that spirit, read on, explore, and enjoy. Precious orchids await.

Anja Klaffenbach

SCHÄTZE ENTDECKEN UND BEWAHREN

Dass es rund 30.000 verschiedene Orchideenarten auf der Welt gibt, hätte ich in meiner Kindheit nie für möglich gehalten: Schließlich begegnete mir in fast jedem Wohnzimmer von Familie und Freunden die immer gleiche *Phalaenopsis*, die für mich lange Jahre ein Synonym für Orchideen war. Ausflüge in verschiedene botanische Gärten, speziell in die wunderschönen Gewächshäuser von Kew Gardens, zeigten mir dann im Laufe der Jahre, wie viele Gattungen und Arten tropischer Orchideen noch darauf warteten, von mir entdeckt zu werden – der Beginn einer wachsenden Faszination.

Wie viele bezaubernde heimische Orchideenarten es in meiner unmittelbaren Nähe gibt, erfuhr ich erst durch Naturschutzaktionen der Arbeitskreise Heimische Orchideen (AHO), die jeweils mit der Kür zur „Orchidee des Jahres" auf besonders gefährdete Arten aufmerksam machen. Vielleicht gelingt es ja auch mit diesem Buch, ein wenig mehr Aufmerksamkeit für den Schutzbedarf vieler so wertvoller Raritäten unserer Natur zu erzeugen.

Viel Freude beim Erkunden und Genießen
von wunderbaren Orchideenschätzen!

Anja Klaffenbach

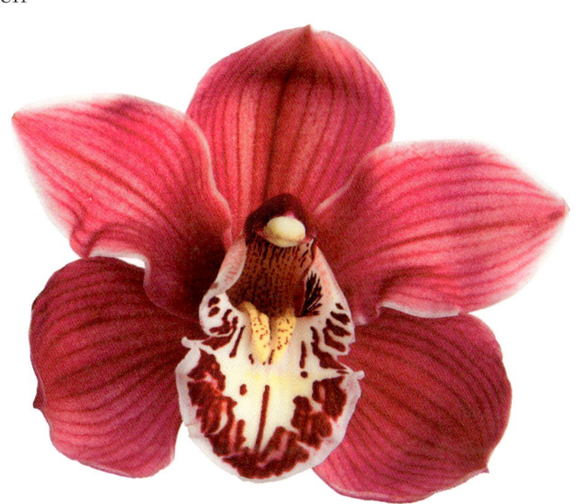

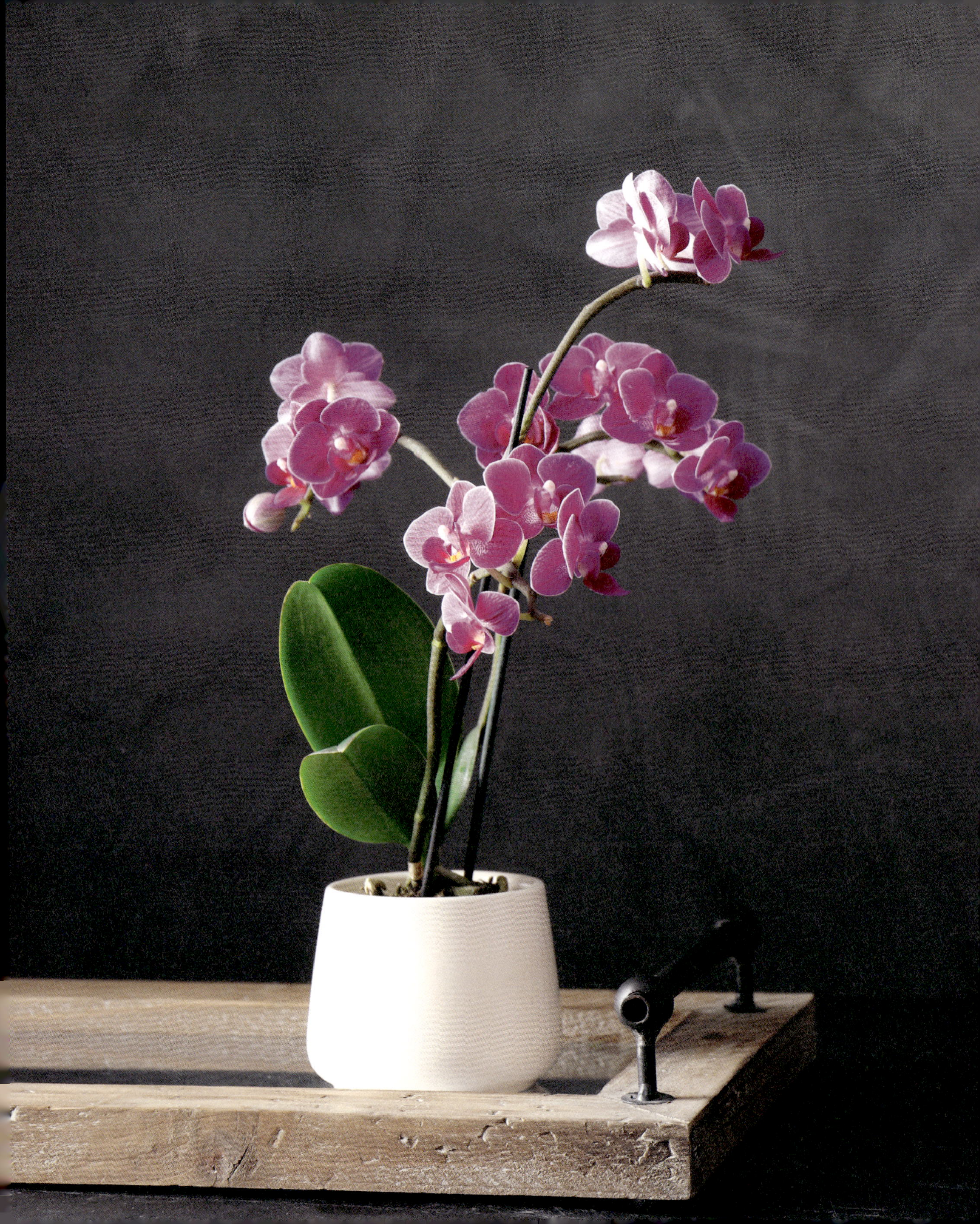

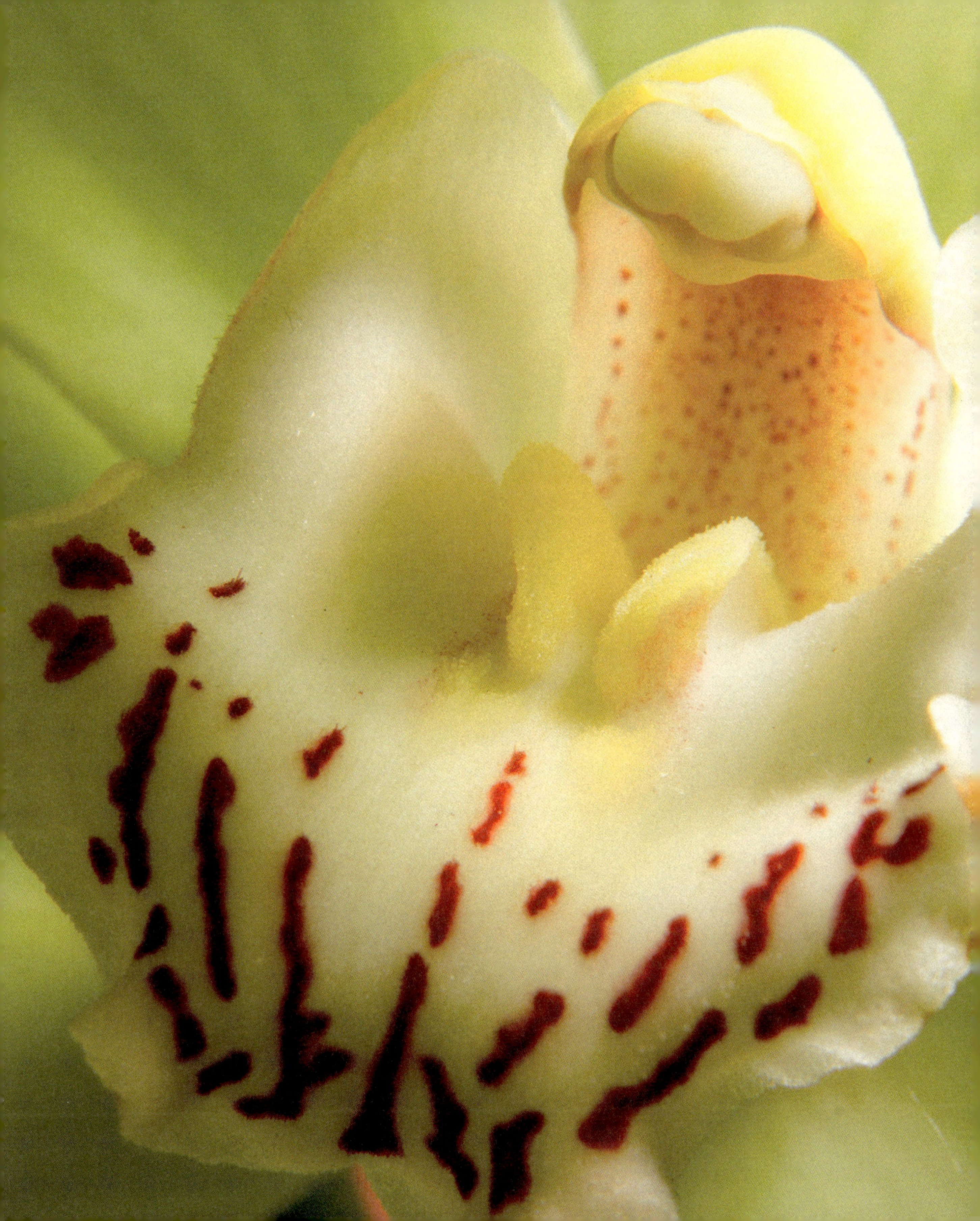

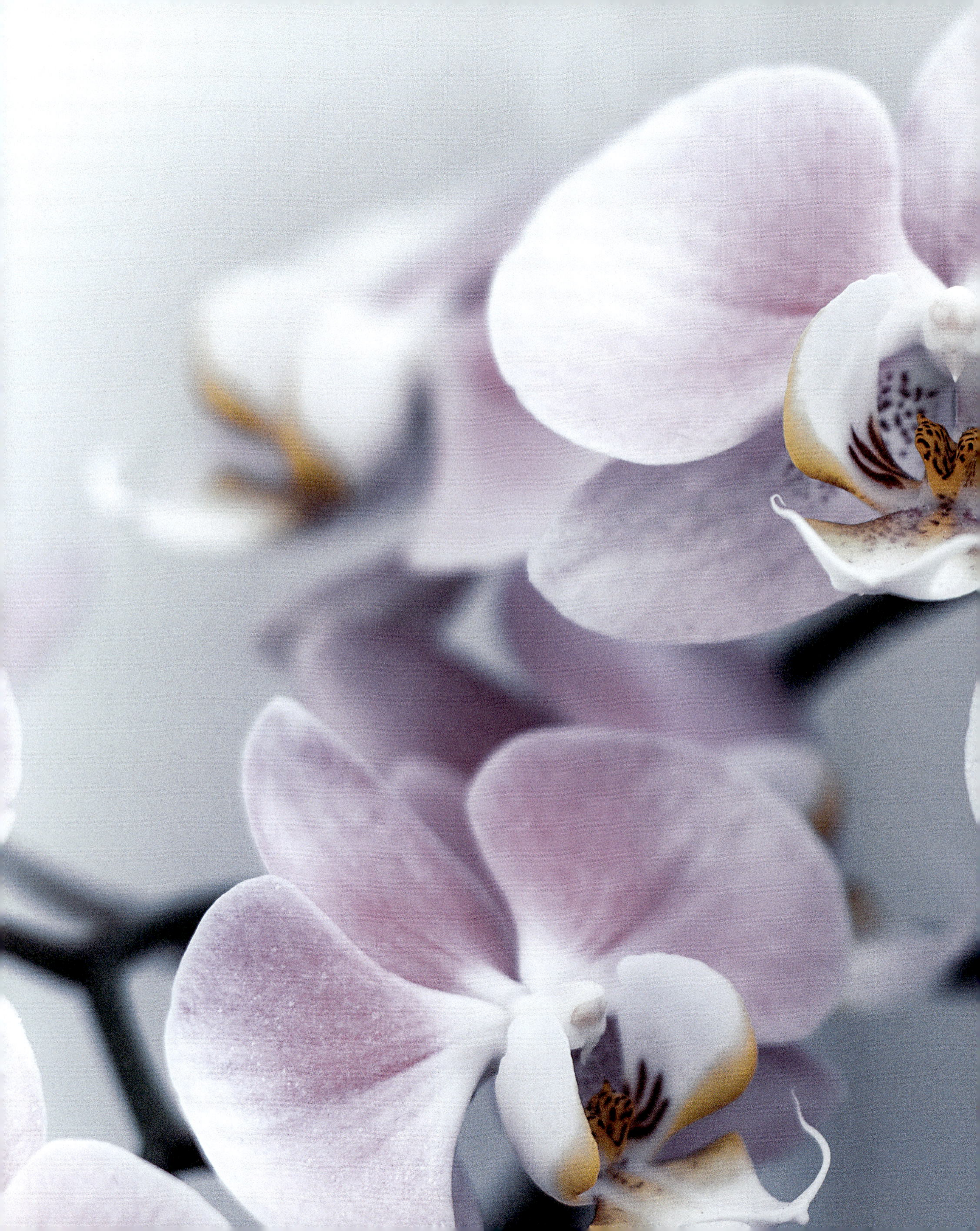

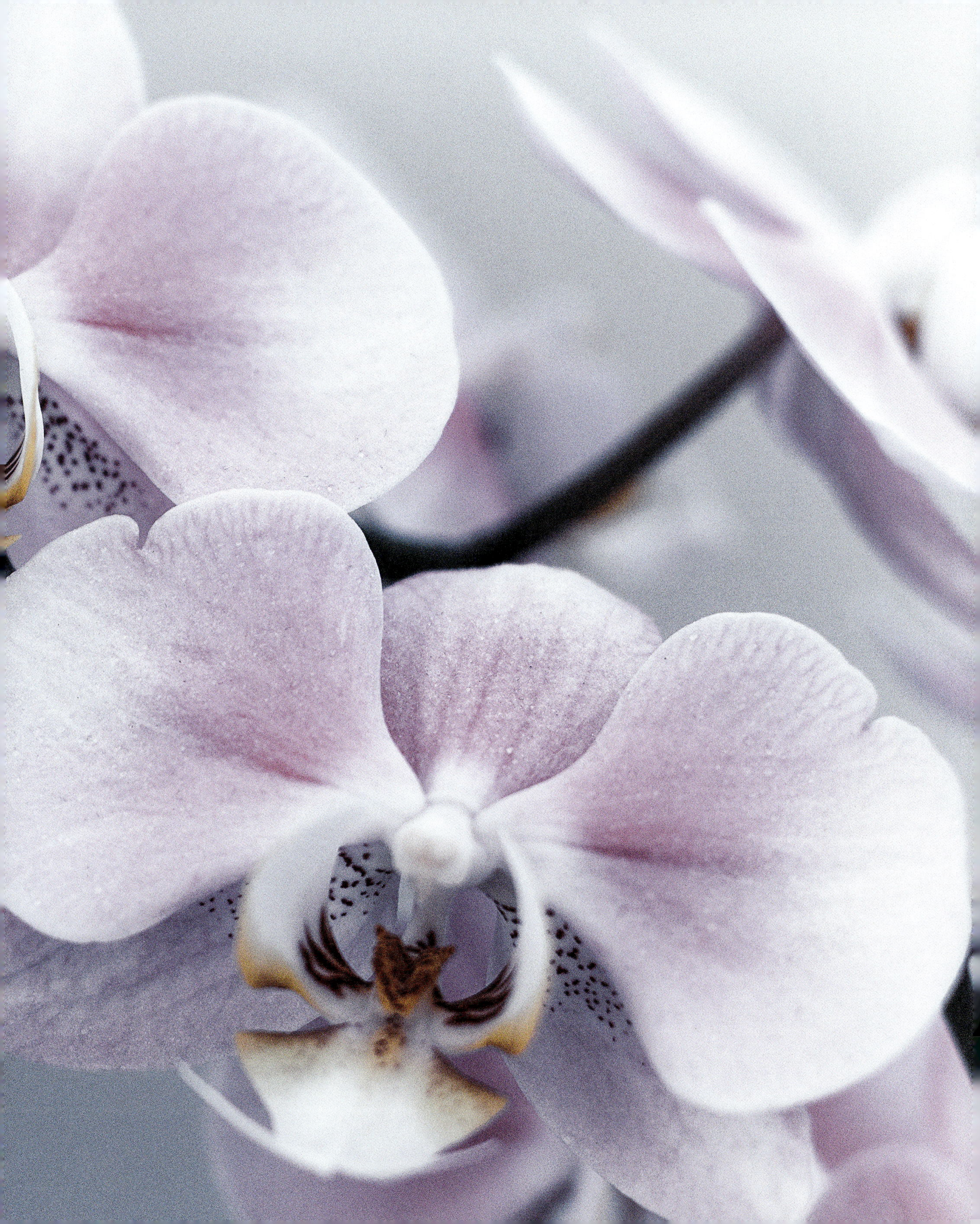

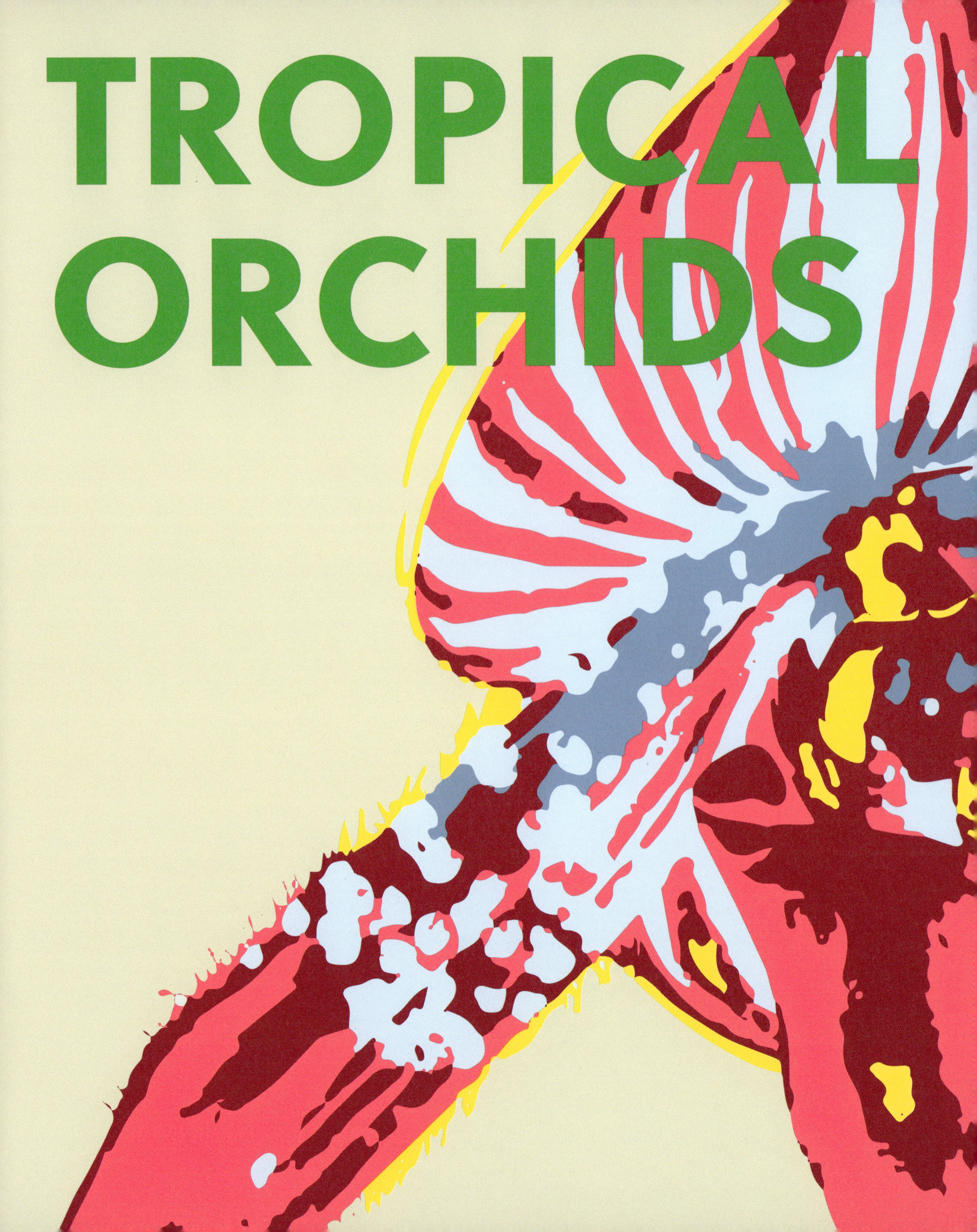
TROPICAL
ORCHIDS

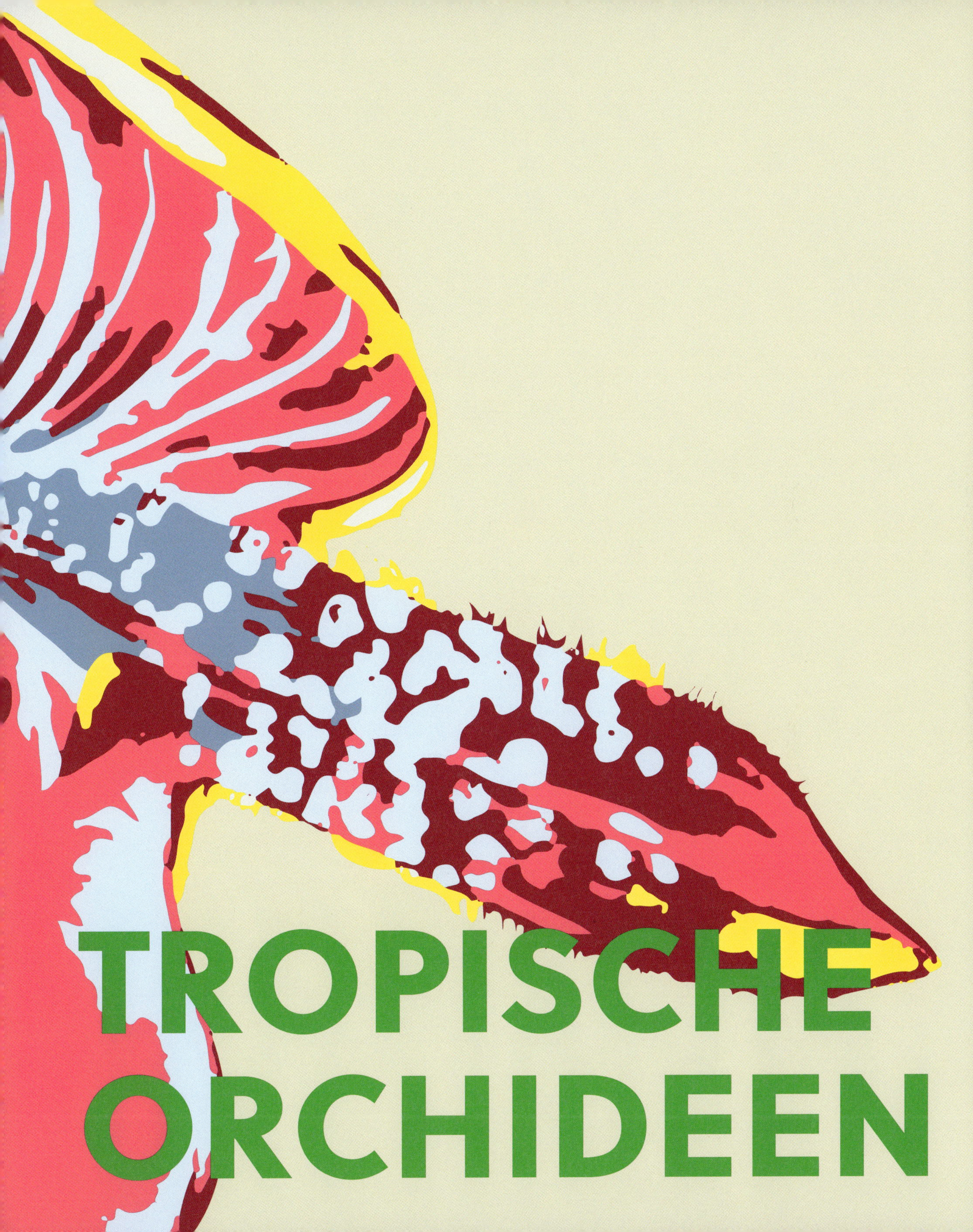

TROPISCHE
ORCHIDEEN

Exotic beauties

THE MAGNIFICENT DIVERSITY OF THE RAINFOREST

THIRTY THOUSAND DIFFERENT ORCHIDS GUARANTEE ORCHID LOVERS COLOR AND VARIETY

The orchids family of plants is thought to have originated in rainforests in what is now Indonesia, and today, around 90% of orchid species are native to the tropics. While they originally grew in soil, orchids have gradually adapted to new habitats. Today, the typical tropical orchid is an epiphyte, growing on trees. This is an evolutionary advantage, as the high canopy provides the perfect balance of light and shade. Rainforest trees are not food for epiphytic orchids; they are simply a habitat. Epiphytic orchids are not parasitic plants that sap the resources of their hosts. But how do these beautiful orchids survive at such lofty heights? The answer lies in their aerial roots, which cling to their rootstock. The silvery, shimmering velamen that surrounds these roots helps them absorb the meager nutrient mix they need to survive. These orchids are content with what they have, obtaining nutrients from the frequent rainfall, high humidity, and decaying plant matter of their host trees. They are true survival artists.

NEXT PAGE | Orchids are native to the tropics.

Exotische Schönheiten

PRÄCHTIGE VIELFALT IM REGENWALD

RUND 30.000 ORCHIDEENARTEN GIBT ES – DAMIT IST ORCHIDEENLIEBHABERN FARBENFROHE ABWECHSLUNG GARANTIERT

Ihren Ursprung hat die Familie der Orchideen im indonesischen Regenwald, und in der Tat sind auch heute rund 90 Prozent der Orchideenarten in den Tropen beheimatet. Waren sie anfänglich im Erdboden verwurzelt, haben die Pflanzen nach und nach neue Lebensräume erobert. Tropische Orchideen wachsen heute in der Regel epiphytisch, also als „Aufsitzer" auf Bäumen. Ein cleverer Schachzug der Evolution, bieten doch diese „Logenplätze" hoch oben in der Nähe der Baumwipfel genau den richtigen Mix aus Licht und Schatten. Die Regenwaldbäume sind dabei lediglich Lebensraum, nicht Nahrung: Epiphytische Orchideen sind keine Schmarotzerpflanzen, die ihre Wirte „anzapfen" und aussaugen. Wovon aber leben die Schönheiten in luftiger Höhe? Einen Hinweis geben die Luftwurzeln, mit denen sich die Orchideen an ihrer Unterlage festhalten: Über das silbrig schimmernde Velamen, das diese umgibt, können sie den für sie passenden kargen Nährstoffmix beziehen. Sie begnügen sich mit den häufigen Regenfällen, der daraus resultierenden hohen Luftfeuchtigkeit und verrottenden Pflanzenresten ihrer „Gastgeberbäume" – echte Überlebenskünstler also.

NÄCHSTE SEITE | Ursprünglich kommen Orchideen aus den Tropen.

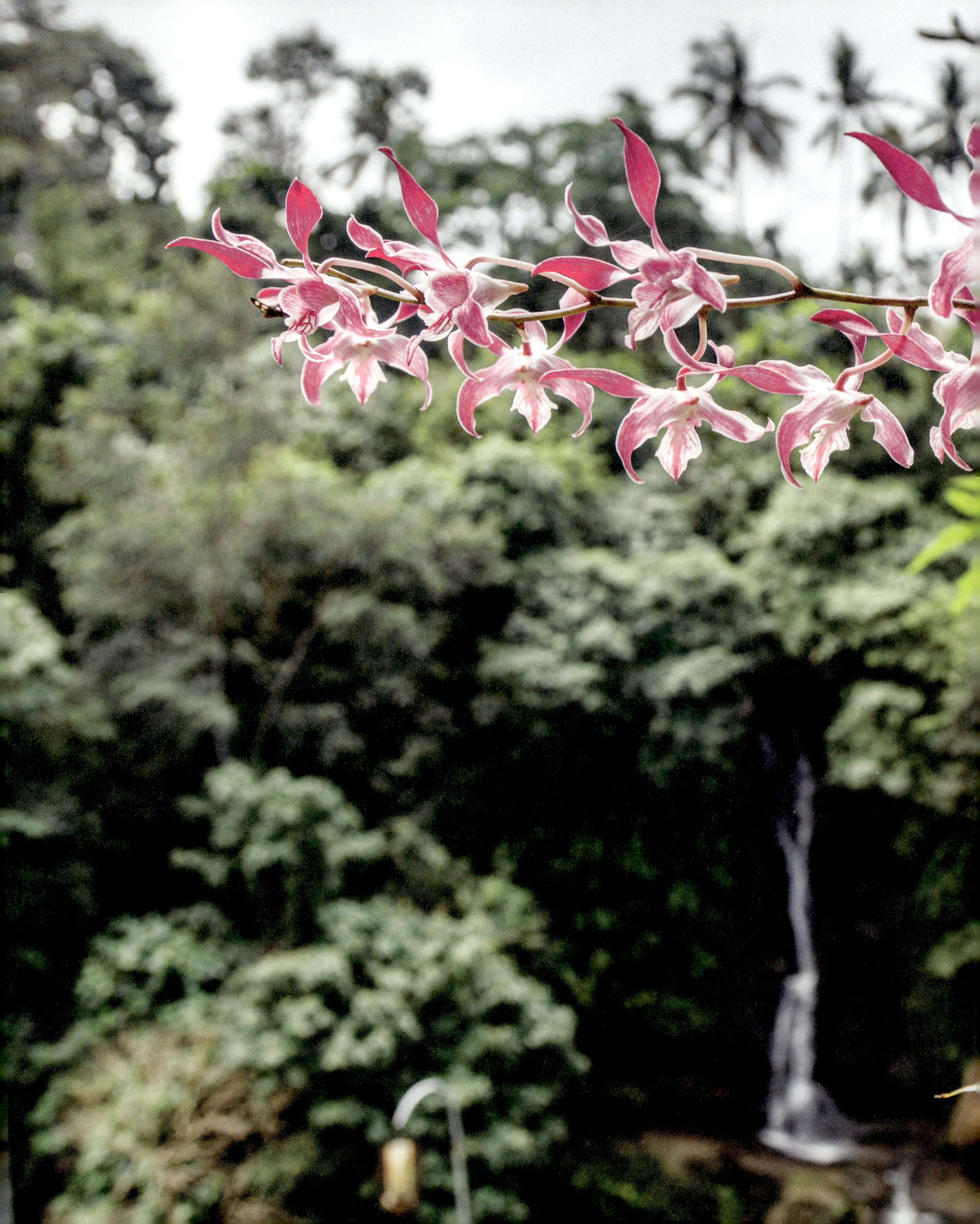

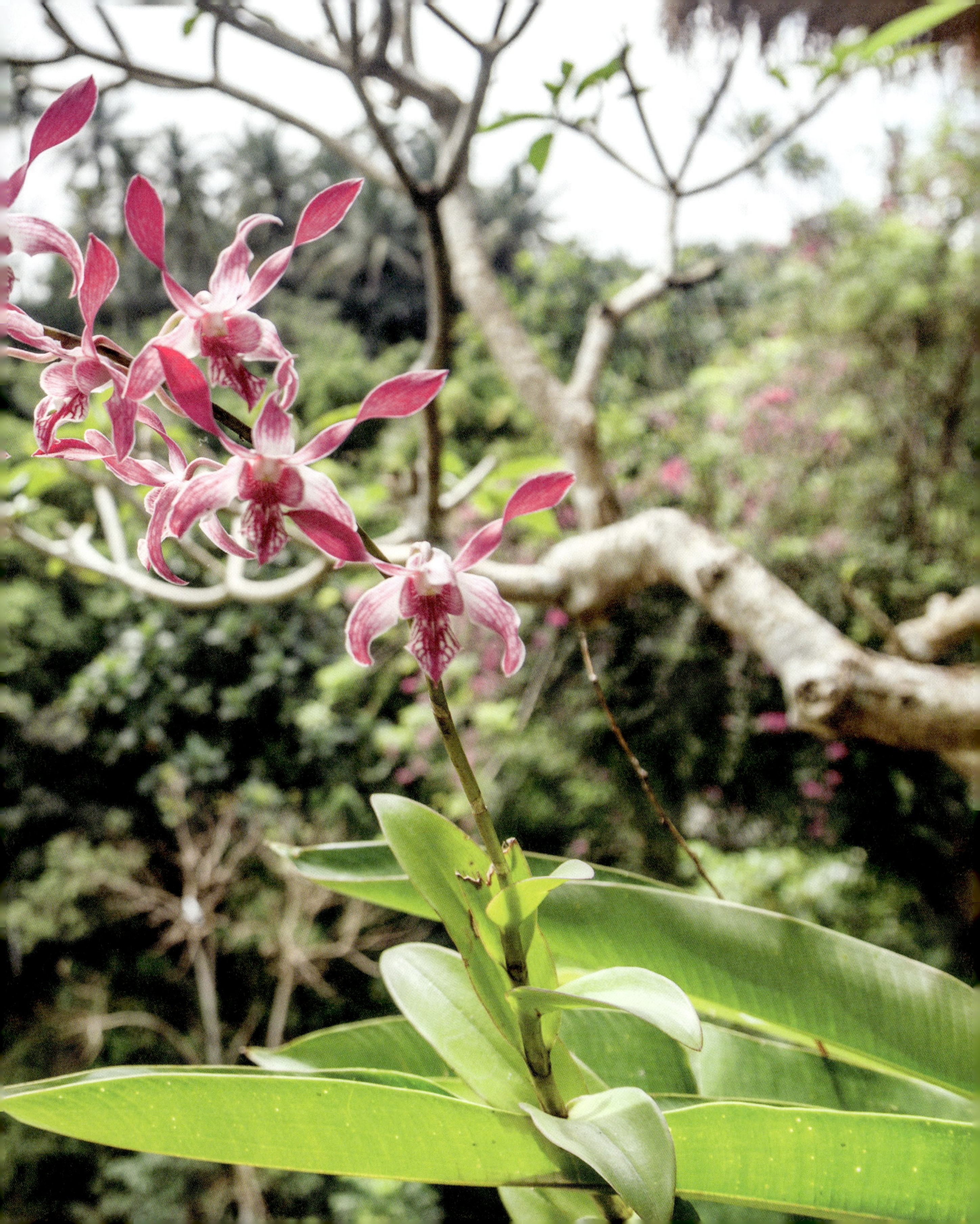

Phalaenopsis

MOTH ORCHID

FROM RARE-AND-EXOTIC TO MASS PRODUCTION

Phalaenopsis orchids are like a gateway drug, wearing down barriers and inhibitions until yet another casual purchaser becomes a fanatical collector. Reams of moth orchids are on sale at low-end supermarkets and garden centers; they've been called "hardware store" orchids – an indignity. Still, they're one of the most popular orchids in the world. These lovely flowers were first discovered in the tropical rainforests of Southeast Asia in 1753. The Swedish botanist Carl von Linné named them *Epidendrum amabile*, which means "lovable." And it's easy to see why he gave them that name. Fans of the world's most bought-and-sold orchid will surely agree.

The *Phalaenopsis* orchid got its current name in 1825 from the German-Dutch botanist Carl Ludwig Blume. He was the one who brought thousands of exotic plant species to Europe from his travels through the Indonesian islands. The story goes that he "rediscovered" the *Phalaenopsis* orchid more or less by accident. He thought he saw moths dancing around a tree trunk, but when he got a closer look, he realized they were actually flowers bobbing in the wind. That's how the *Phalaenopsis* orchid got its name: Ancient Greek *phalaina*, "moth-like". There are so many different species of Phalaenopsis orchids, it's hard to keep track of them all. And so the cultivated hybrids that keep coming to market mostly remain nameless.

Phalaenopsis

SCHMETTERLINGS-ORCHIDEE

VON DER EXOTISCHEN SELTENHEIT ZUM MASSENPRODUKT

Für viele Orchideenliebhaber ist die *Phalaenopsis* quasi die „Einstiegs-droge" in ihre Sammelleidenschaft. Heute ist die Schmetterlingsorchidee auch bei Discountern und Gartencentern in rauen Mengen im Angebot, was ihr den etwas despektierlichen Spitznamen „Baumarktorchidee" eintrug. Ihrer Beliebtheit tut das indes keinen Abbruch. 1753 wurde die Blühschönheit in den tropischen Regenwäldern Südostasiens entdeckt und vom schwedischen Botaniker Carl von Linné als *Epidendrum amabile* beschrieben – und das Epitheton *amabilis*, also „liebenswert", bestätigen auch heute noch die Fans der wohl meistverkauften Orchidee weltweit.

Ihren heutigen Gattungsnamen erhielt die *Phalaenopsis* 1825 vom deutsch-niederländischen Botaniker Carl Ludwig Blume, der von seinen Reisen durch die indonesische Inselwelt tausende exotischer Pflanzenarten nach Europa brachte. Man erzählt sich, dass er die Orchidee mehr oder weniger durch Zufall „wiederentdeckte": Er glaubte, Nachtfalter (auf Griechisch *phalaina*) um einen Baumstamm tanzen zu sehen. Bei näherer Betrachtung entpuppten diese sich als im Wind wippende Blüten: Der Gattungsname *Phalaenopsis* war geboren. Wie viele Arten der *Phalaenopsis* es heute gibt, ist kaum mehr zu überblicken, und so bleiben die immer neuen Kulturhybriden für den Massenmarkt meist namenlos.

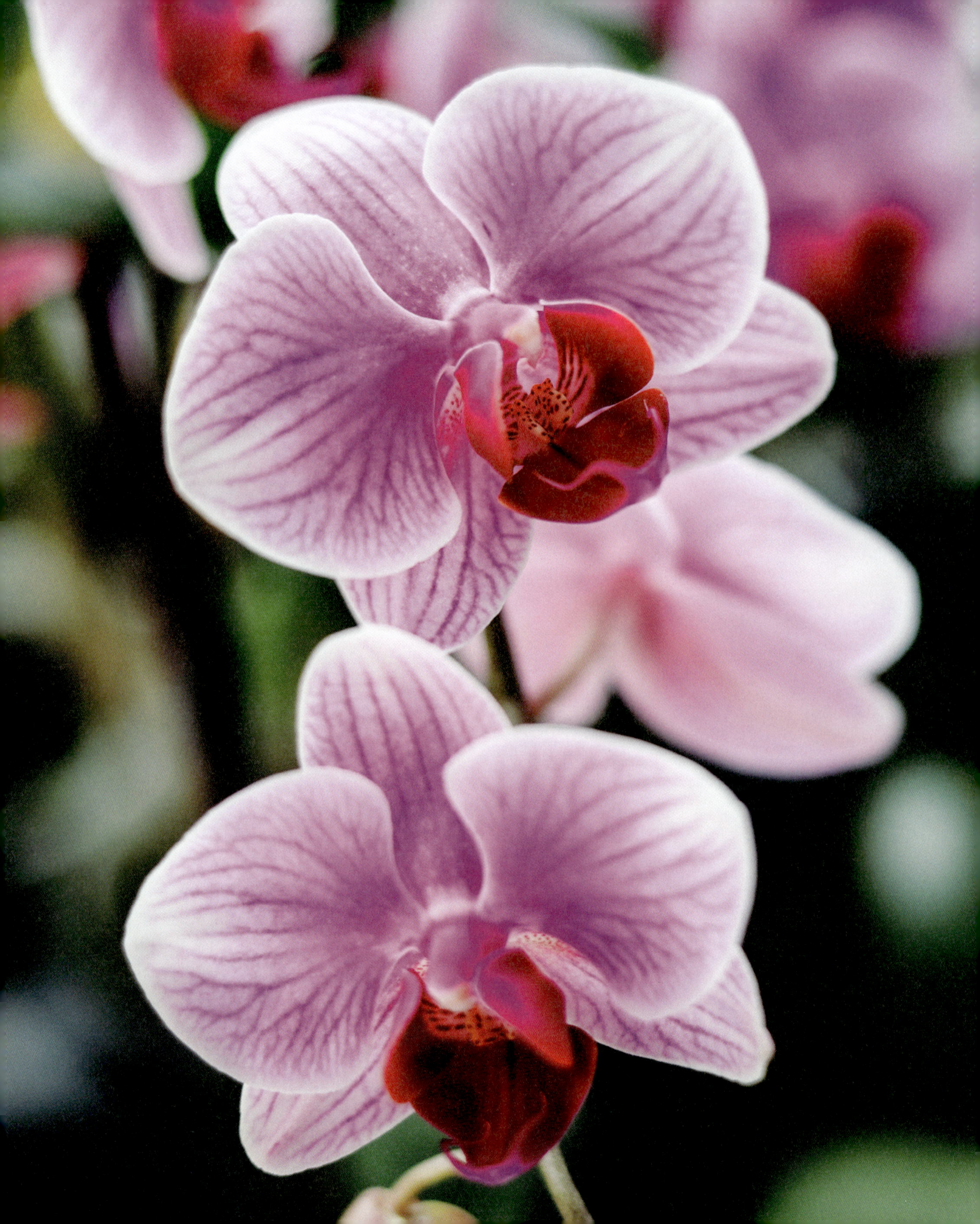

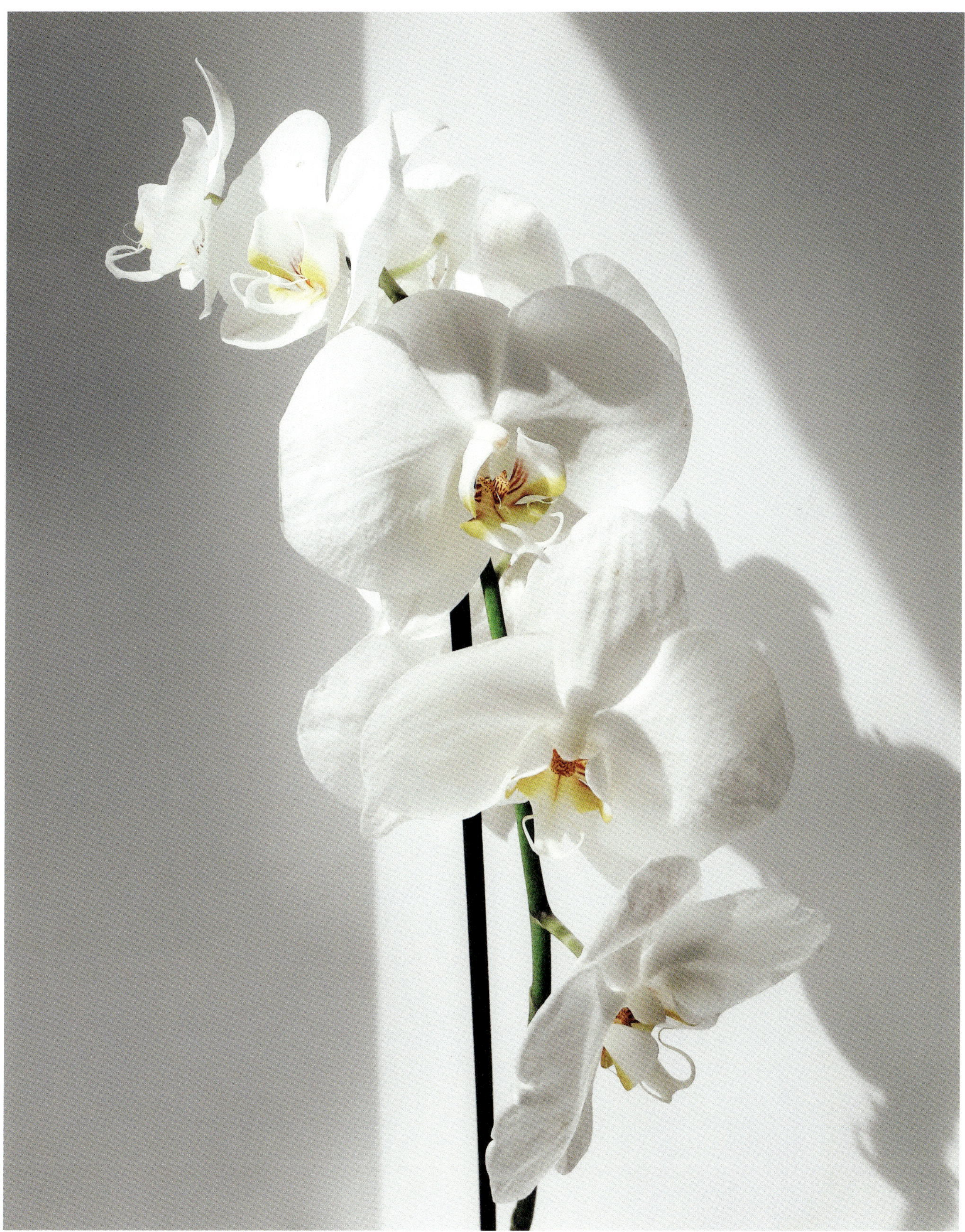

Brassia

SPIDER ORCHID

ELEGANT AND EFFICIENT

It's unclear whether arachnophobes will be thrilled by the *Brassia*, a tropical Central American orchid, due to its long, narrow petals that bear a striking resemblance to spider legs. However, these delicate flowers serve a practical purpose for the orchid, as some *Brassia* species attract spider wasps for pollination. These wasps, in turn, use spiders as hosts for their larvae. Additionally, since the *Brassia's* intricate blooms grow in abundant clusters, the plant may be more alluring than a lone host animal.

SPINNENORCHIDEE

ELEGANT UND EFFIZIENT

Ob Arachnophobiker sich für die aus dem tropischen Mittelamerika stammende *Brassia* begeistern können, sei dahingestellt: Die langen, schmalen Blütenblätter erinnern doch sehr an die Beine einer Spinne. Praktisch aber für die Orchidee selbst, denn zum Bestäuben locken einige *Brassia*-Arten Wegwespen an, die ihrerseits Spinnen als Wirte für ihre Larven suchen. Da die filigranen Blüten der *Brassia* in üppigen Trauben wachsen, wirkt die Orchidee vielleicht erfolgversprechender als ein einzelnes Tierchen.

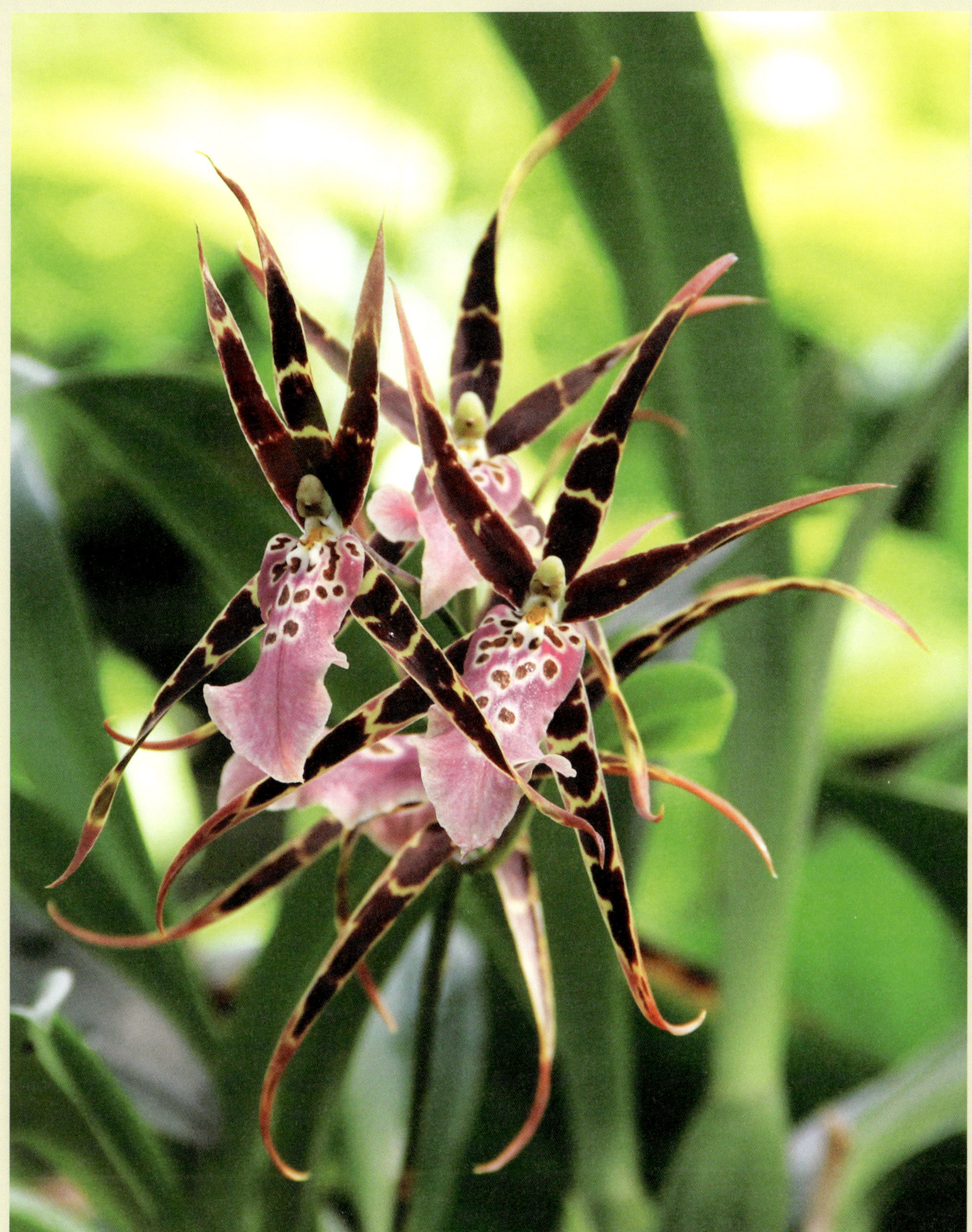

Paphiopedilum

VENUS SLIPPER

CAREFUL, IT'S A DOUBLE TRAP

With the shoe-shaped lip of its flower, the *Paphiopedilum* lures insects into its trap for pollination. The tropical Venus slipper also poses a challenge for orchid enthusiasts, as its bloom is nearly identical to another slipper orchid, *Cypripedium*. One way to distinguish the two is by their blooming periods: while the *Cypripedium*, which is native to northern regions, only blooms in early summer, the *Paphiopedilum*, often kept as a houseplant, shows its full glory mostly in winter.

VENUSSCHUH

VORSICHT, FALLE – IM DOPPELTEN SINN

Mit ihrer schuhförmigen Blütenlippe lockt die *Paphiopedilum* Insekten zur Bestäubung in die Falle. Eine Falle stellt der tropische Venusschuh auch Orchideenfreunden, denn die Blüte sieht dem Frauenschuh *Cypripedium* zum Verwechseln ähnlich. Eine Unterscheidungshilfe ist die Blütezeit: Während die aus nördlichen Gefilden stammende *Cypripedium* nur im Frühsommer blüht, zeigt die *Paphiopedilum* als Zimmerpflanze ihre Pracht meist im Winter.

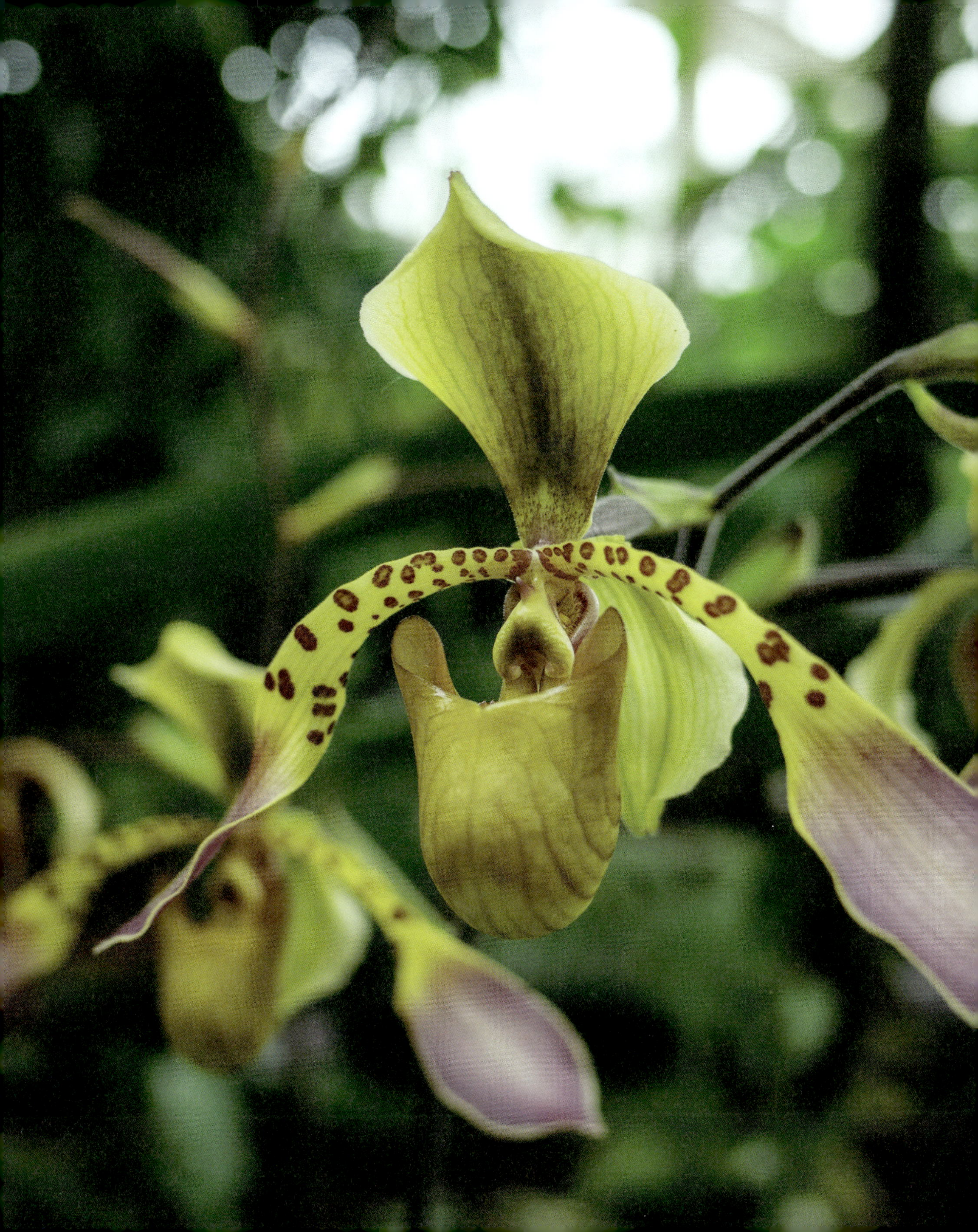

Dendrobium nobile

NOBLE DENDROBIUM

A BEAUTY WITH A MEDICAL CAREER

Its name gives it away: *Dendrobium nobile* lives in nature on trees, or in Greek, *dendron*. It is one of five hundred species that grow in Asia, and is the emblem of the old kingdom of Sikkim, today an Indian province. *Dendrobium nobile* also plays a role in medicine: the orchid is one of the fifty most important plants in traditional Chinese medicine.

TRAUBENORCHIDEE

EINE SCHÖNHEIT MIT MEDIZINISCHER KARRIERE

Ihr Name verrät es: *Dendrobium nobile* lebt in der Natur auf Bäumen, griechisch *dendron*. Die Blüten der in Asien beheimateten Orchideengattung mit weit über 1000 Arten wachsen eng gruppiert direkt aus dem Stamm, woher ihr Trivialname Traubenorchidee rührt. Und auch in der Heilkunde spielt *Dendrobium nobile* eine Rolle: Die Orchidee zählt zu den 50 wichtigsten Pflanzen der Traditionellen Chinesischen Medizin.

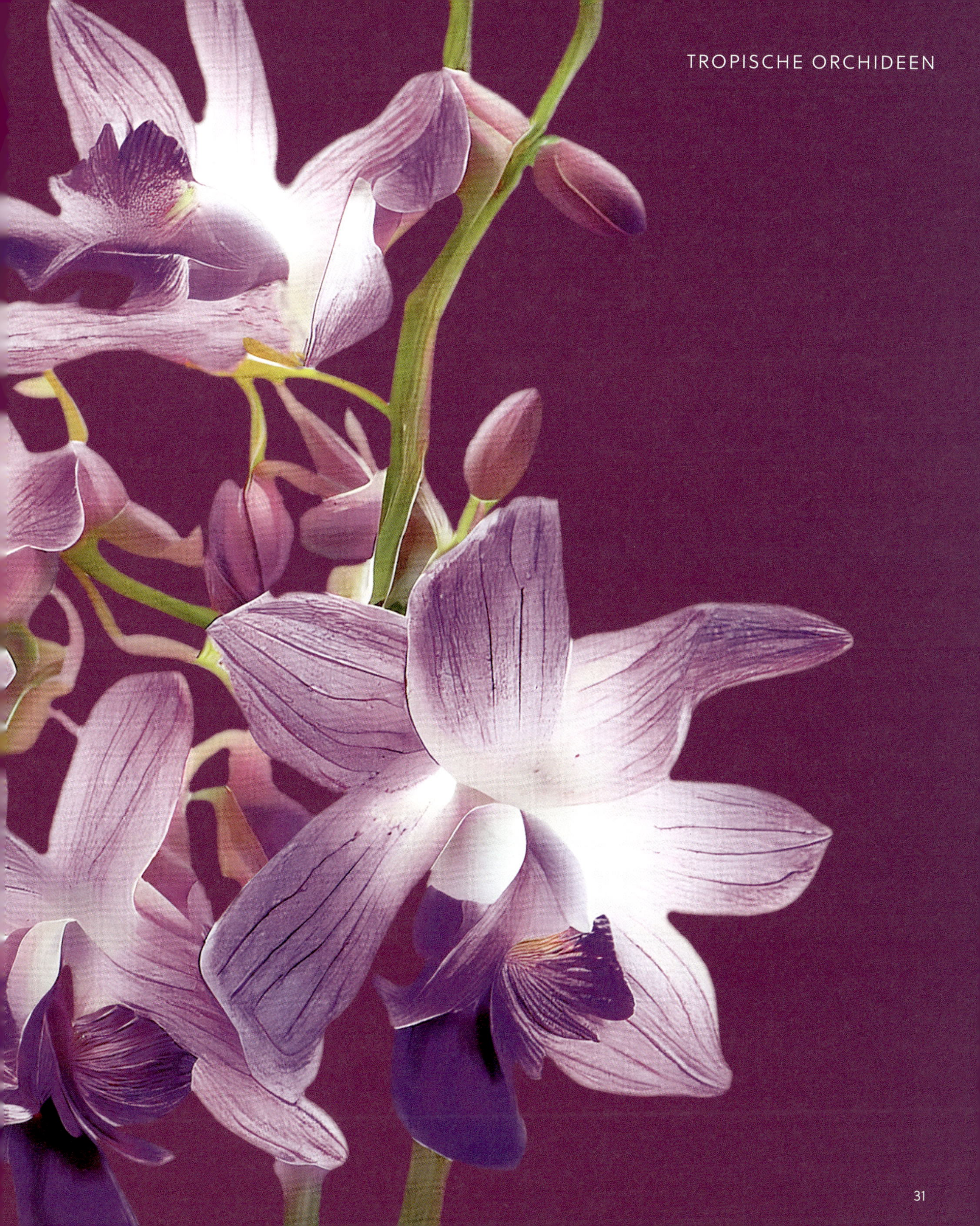

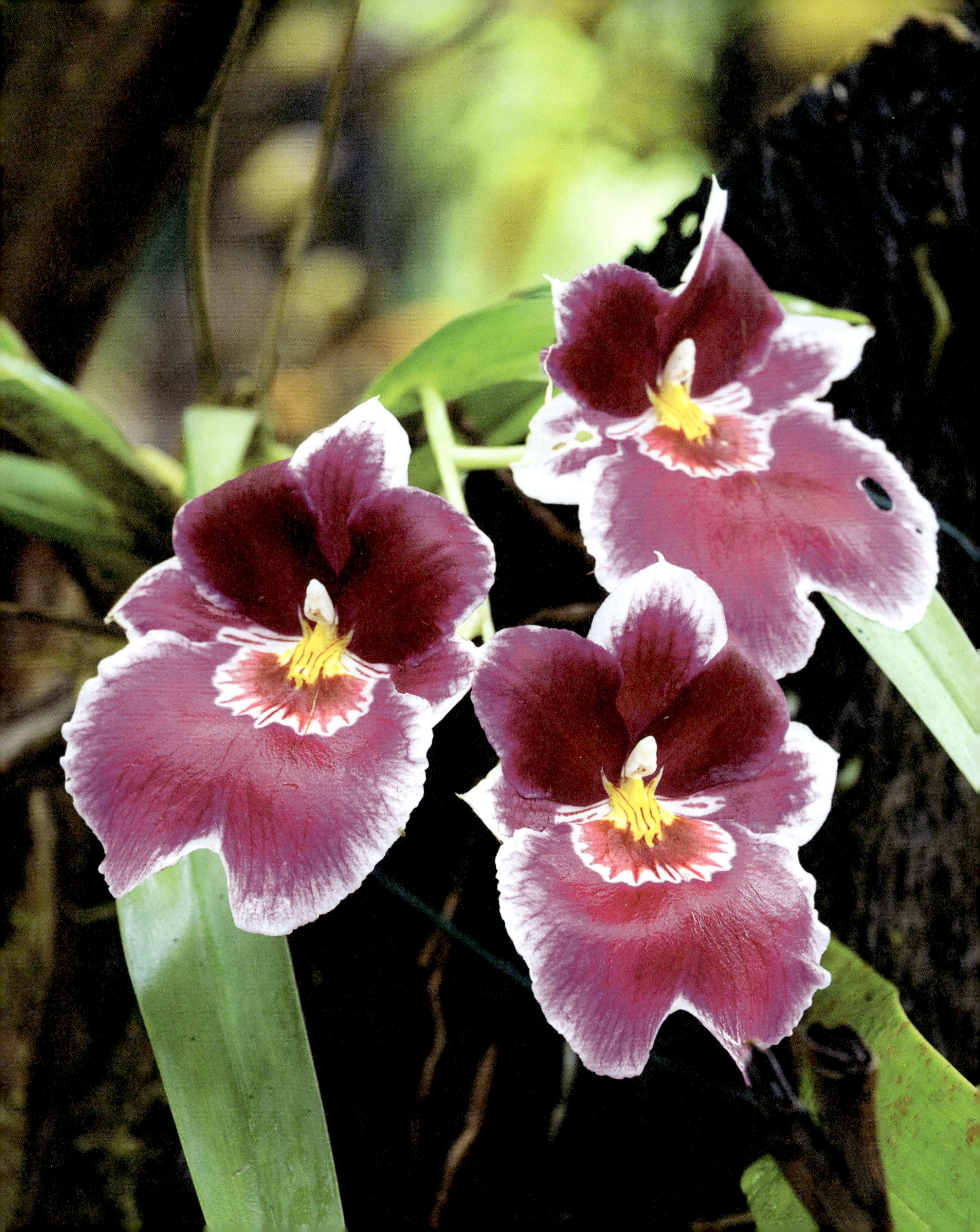

Miltoniopsis

PANSY ORCHID

DRAMATIC COLORS AND FRAGRANCE

The *Miltoniopsis* orchid, native to the mountain forests of the northern Andes, is a stunning sight. Its flowers, with their distinctive eye-like patterns and rounded shape, have earned it the nickname "pansy orchid." Some people mistake it for the Brazilian genus *Miltonia*, but only those familiar with their distinct requirements will have success with these false twins. While cool and damp conditions are ideal for *Miltoniopsis*, *Miltonia* can tolerate a warmer and drier environment.

STIEFMÜTTERCHEN-ORCHIDEE

FARBENPRÄCHTIG UND WOHLRIECHEND

Die schöne *Miltoniopsis* hat ihre Heimat in den Bergwäldern der nördlichen Anden. Ihren Blüten mit dem auffälligen Auge und der rundlichen Form verdankt sie die Bezeichnung Stiefmütterchen-Orchidee. Oft wird sie mit der Gattung *Miltonia* aus Brasilien verwechselt – doch nur, wer die unterschiedlichen Ansprüche der falschen Zwillinge kennt, wird mit ihnen glücklich werden: Während *Miltoniopsis* es eher kühl und feucht mag, toleriert die *Miltonia* auch eine wärmere, trockenere Umgebung.

Zygopetalum

"Z."

AN ORCHID THAT WEARS A CROWN

Zygopetalum – genus name *Z.* – is a striking orchid that grows on tree trunks in the Brazilian rainforest, or, often, on the forest floor. Its Greek name, *zygon*, or "yoke," may not quite be it; the magnificent flower looks more like a head that wears an incredible, regal crown, its five petals with colorful spots encircling the likewise beautifully colored labellum. This elegant orchid is also highly prized as a cut flower due to its durable blooms and intense hyacinth-like fragrance.

JOCHKRONE

DIE GEKRÖNTE ORCHIDEE

Im brasilianischen Regenwald wächst *Zygopetalum* auf Baumstämmen, oft auch im Waldboden. Der nicht besonders gebräuchliche Trivialname Jochkrone leitet sich vom griechischen *zygon* – für Joch – ab. Ein „gekröntes Haupt" ist die *Zygopetalum* aber allein schon durch ihre wahrhaft königliche Blüte: Fünf bunt gefleckte Petalen umgeben die wunderschön gefärbte Blütenlippe wie eine Krone. Auch als Schnittblume ist die elegante *Zygopetalum* sehr beliebt: Sie ist lange haltbar und duftet intensiv nach Hyazinthen.

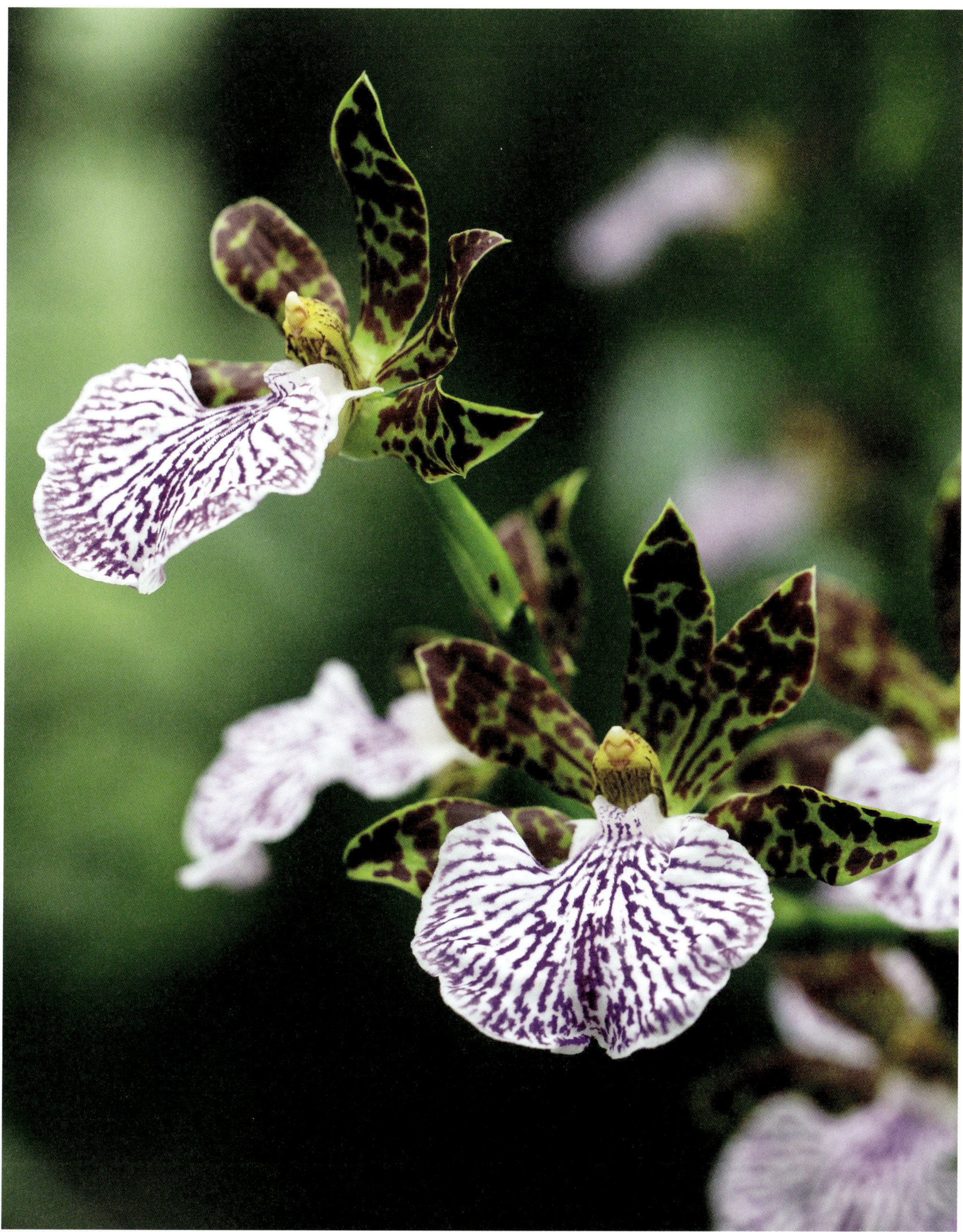

Cattleya

CORSAGE ORCHID

THE ORCHID FOR YOUR NECKLINE

When a *Cattleya* orchid came to Europe in 1818, it may have been by accident, packed in with a shipment of other goods from Brazil, just this small, green tendril among them. But it was not simply filler material *(see page 154)*. When their namesake – English merchant and botanist John Cattley – opened the cargo box, he found that the orchid was blooming beautifully despite the inhospitable environment. The flowers were so fragrant that they became popular as ornaments on the evening gowns of distinguished ladies. In England, they became known as corsage orchids.

KAPUZENORCHIDEE

DIE ORCHIDEE FÜRS DEKOLLETÉ

Die *Cattleya* kam 1818 eher zufällig nach Europa: Als unprätentiöse grüne Ranke wurde die Pflanze fast beiläufig einer Warenlieferung aus Brasilien beigepackt; dass sie nur Füllmaterial gewesen sein soll, gilt heute allerdings als widerlegt *(siehe Seite 155)*. Den widrigen Umständen zum Trotz blühte die Orchidee wunderschön, als der englische Kaufmann und Botaniker John Cattley die Frachtkiste öffnete. Die intensiv duftenden Blüten waren fortan als Schmuck an den Abendroben vornehmer Damen beliebt und wurden in England so als *corsage orchids* bekannt.

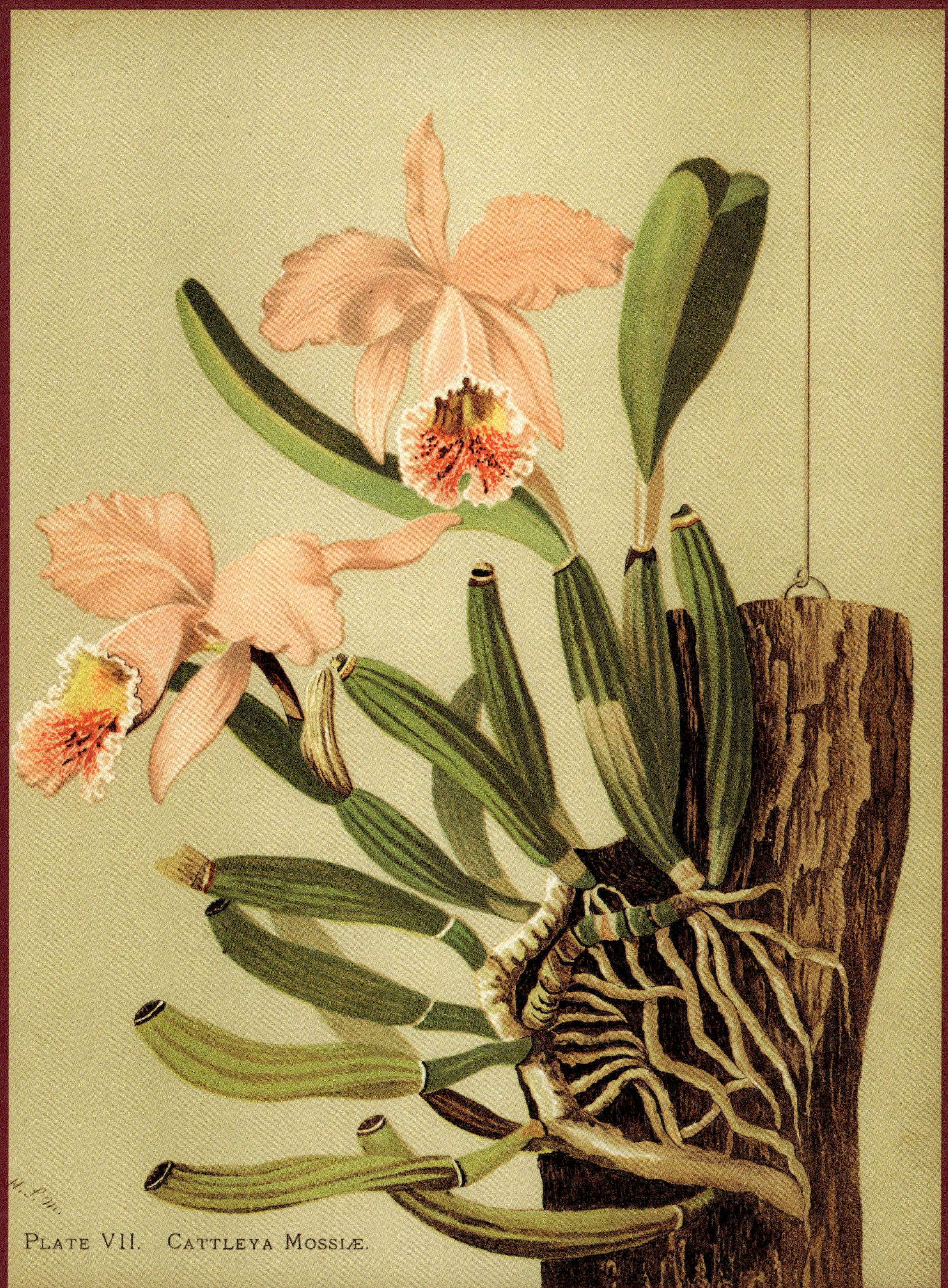

Plate VII. Cattleya Mossiæ.

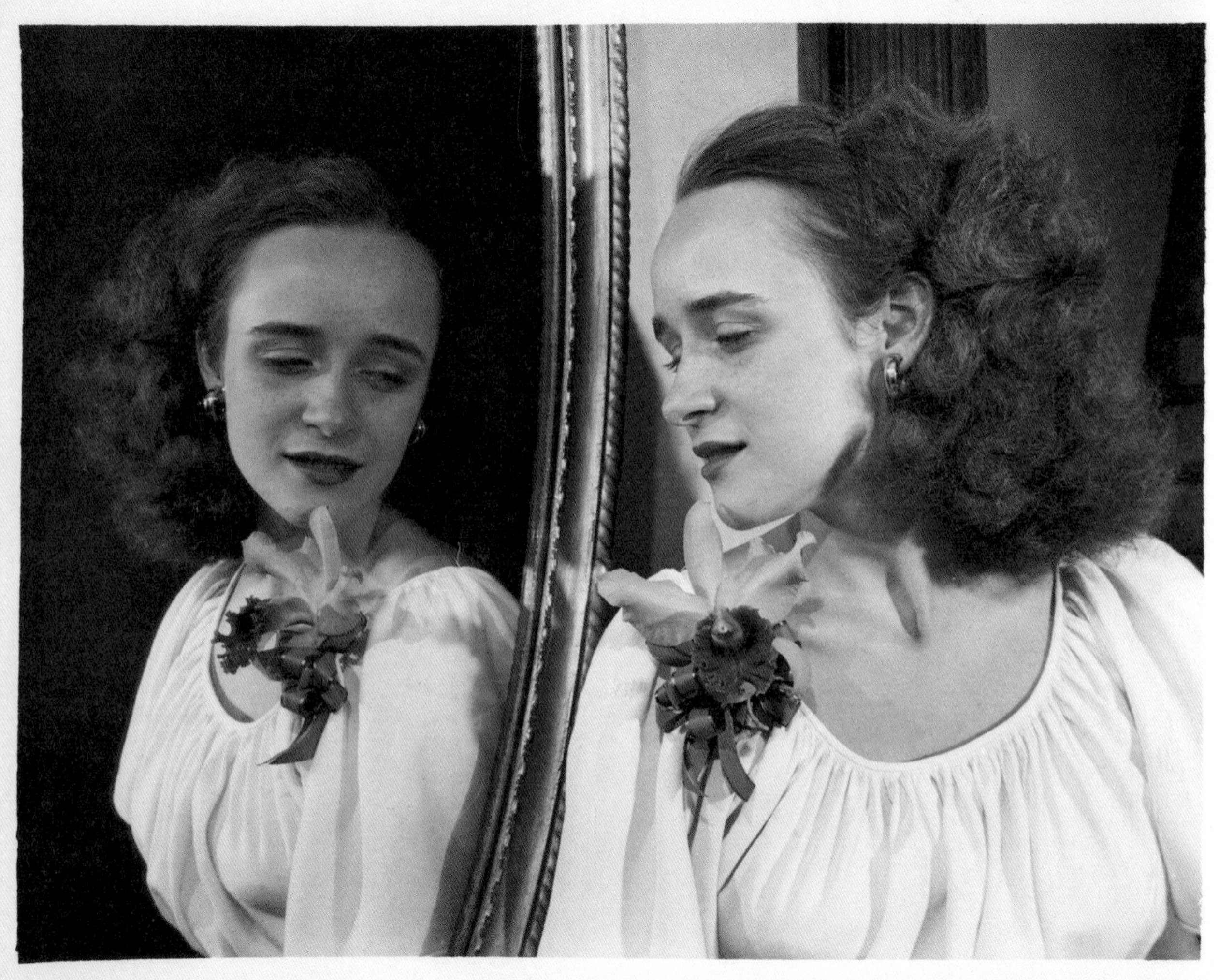

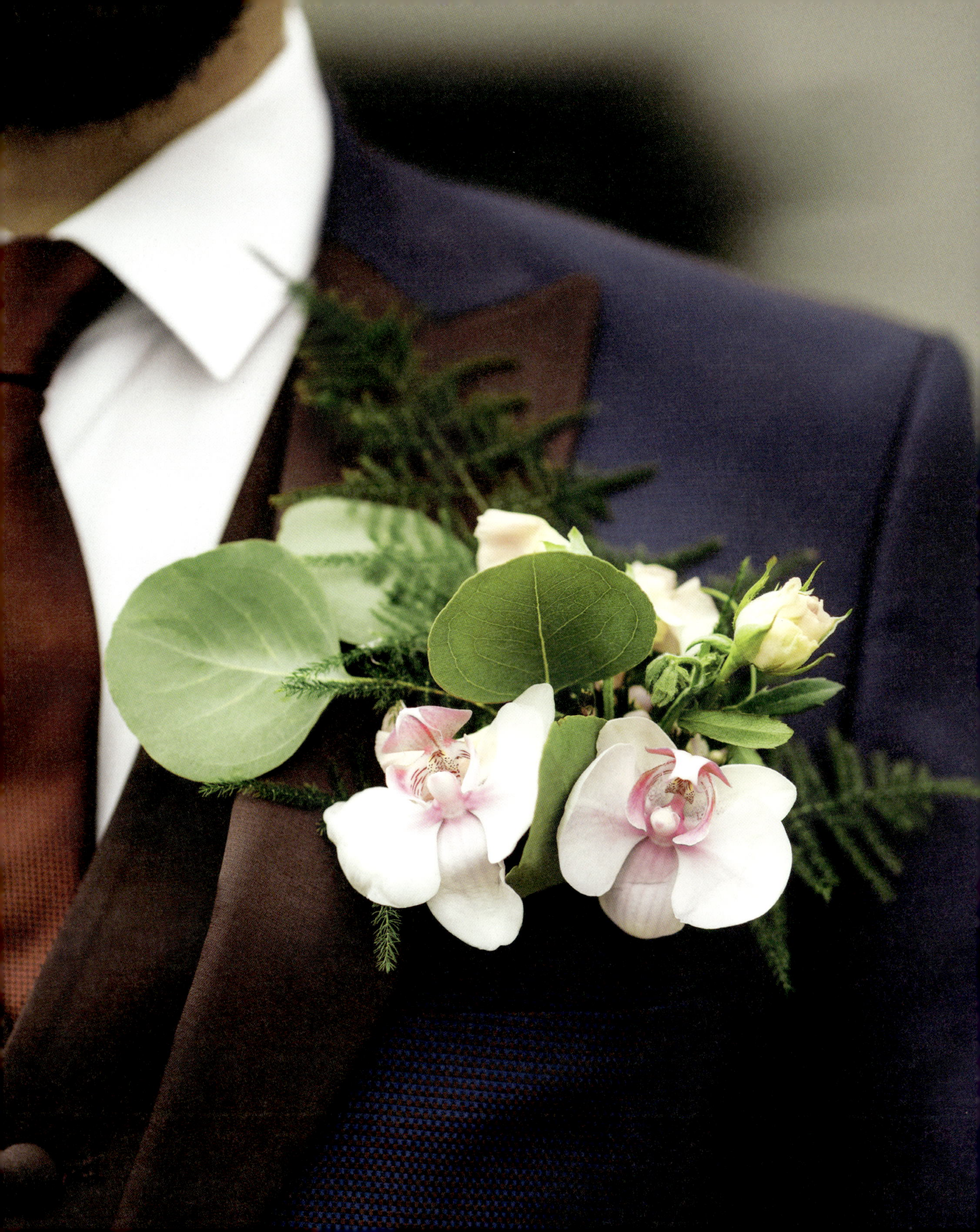

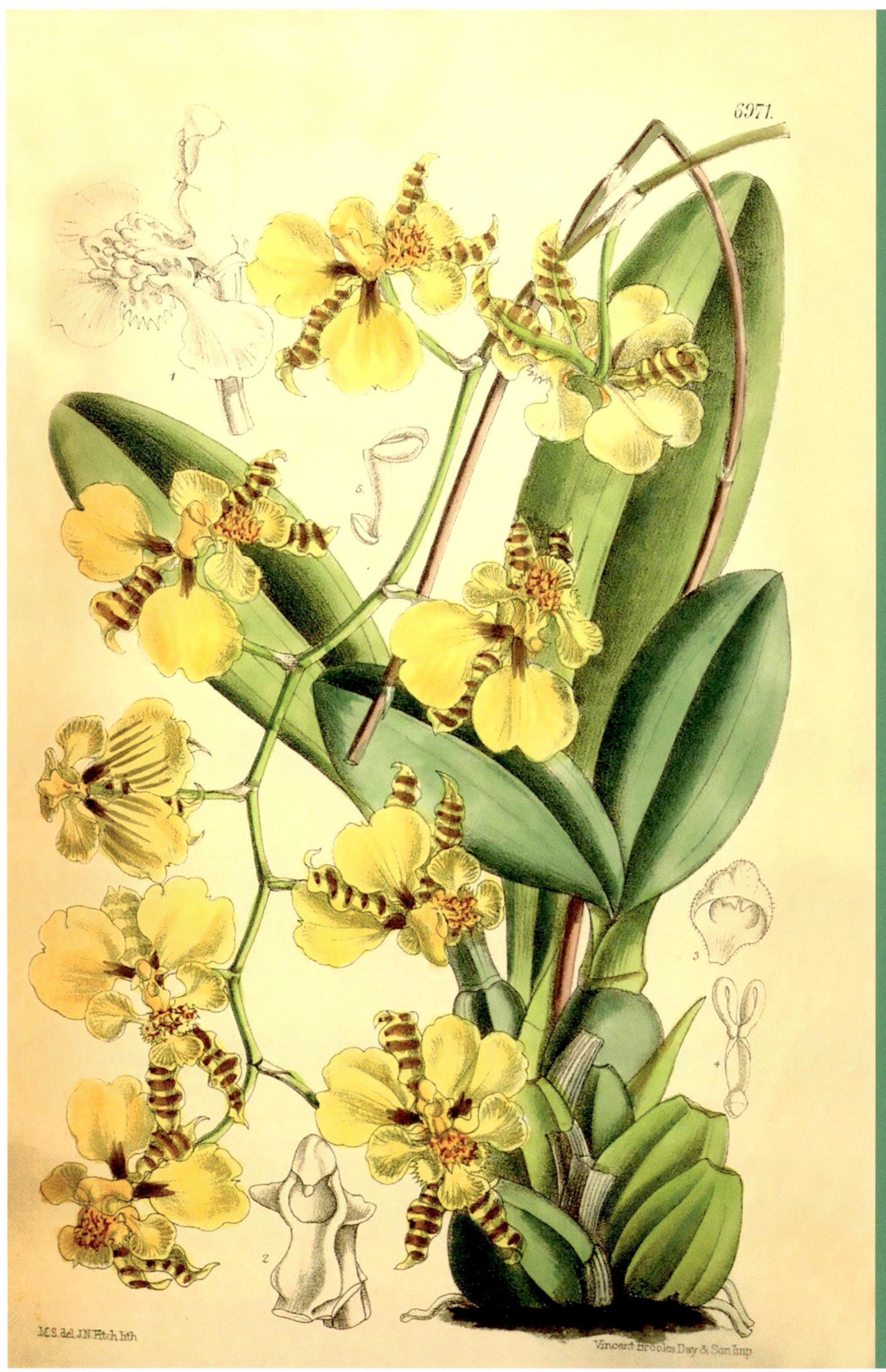

6971.
M.S. del. J.N.Fitch lith.
Vincent Brooks Day & Son Imp

Oncidium

DANCING LADY ORCHID

TIGER OR DANCER?

The flowers of this beautiful orchid dance through the tropical and subtropical forests of South and Central America. The petals of various species of *Oncidium* are mostly yellow-brown striped (which earned it the German nickname *Tigerorchidee*, "tiger orchid"). English orchid lovers were probably more fascinated by the strongly articulated labellum, which reminded them of a magnificent, wide-swinging ball gown – hence the common name, dancing lady orchid.

TIGERORCHIDEE

TIGER ODER TÄNZERIN?

In den tropischen und subtropischen Waldgebieten Süd- und Mittelamerikas lässt diese Orchideenschönheit ihre Blüten tanzen. Die Petalen der *Oncidium*-Arten sind meist gelb-braun gestreift, was ihr den deutschen Beinamen Tigerorchidee einbrachte. Englische Orchideenfreunde waren wohl mehr von den stark ausgeprägten Blütenlippen fasziniert und an ein prachtvolles, weit schwingendes Ballkleid erinnert – so kam dort der Trivialname *Dancing lady orchid* zustande.

Vanda

WIND ORCHID

THE FLOWER OF THE SAMURAI

Vanda, a flower that thrives on love and air, is native to Southeast Asia and the Pacific region. It has distinctive aerial roots that wrap around trees, allowing it to thrive in environments with lots of light, high humidity, and a constant breeze. This is why it is also known as the "wind orchid." *Vanda* blooms in a variety of colors, spanning the entire rainbow spectrum. It has been cultivated in Japan since the seventeenth century and was particularly revered by Samurai warriors, who believed that tying the flowers of *Vanda falcata* to their swords would bring them good luck and help them emerge victorious in battle.

WINDORCHIDEE

DIE BLUME DER SAMURAI

Von Luft und Liebe lebt *Vanda*, die sich von Südostasien bis zum Pazifik mit ihren ausgeprägten Luftwurzeln um Bäume rankt und auf diese Weise findet, was sie braucht: viel Licht, hohe Luftfeuchtigkeit und immer eine frische Brise – daher auch die Bezeichnung Windorchidee. Sie blüht in allen Farben des Regenbogens und wurde schon im 17. Jahrhundert in Japan kultiviert. Samurai-Krieger banden sich Blüten der *Vanda falcata* als Glücksbringer an ihre Schwerter, um den nächsten Kampf siegreich zu überstehen.

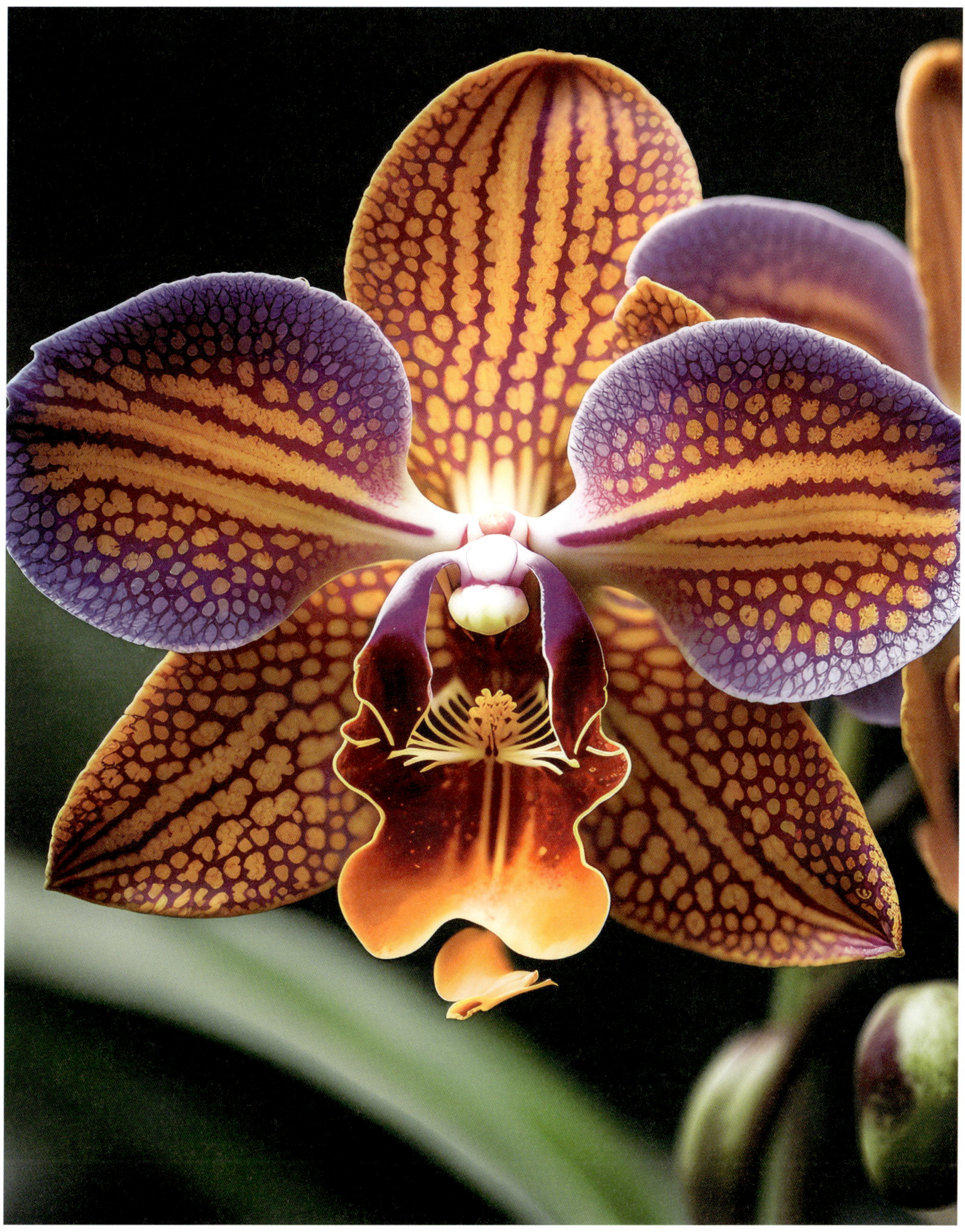

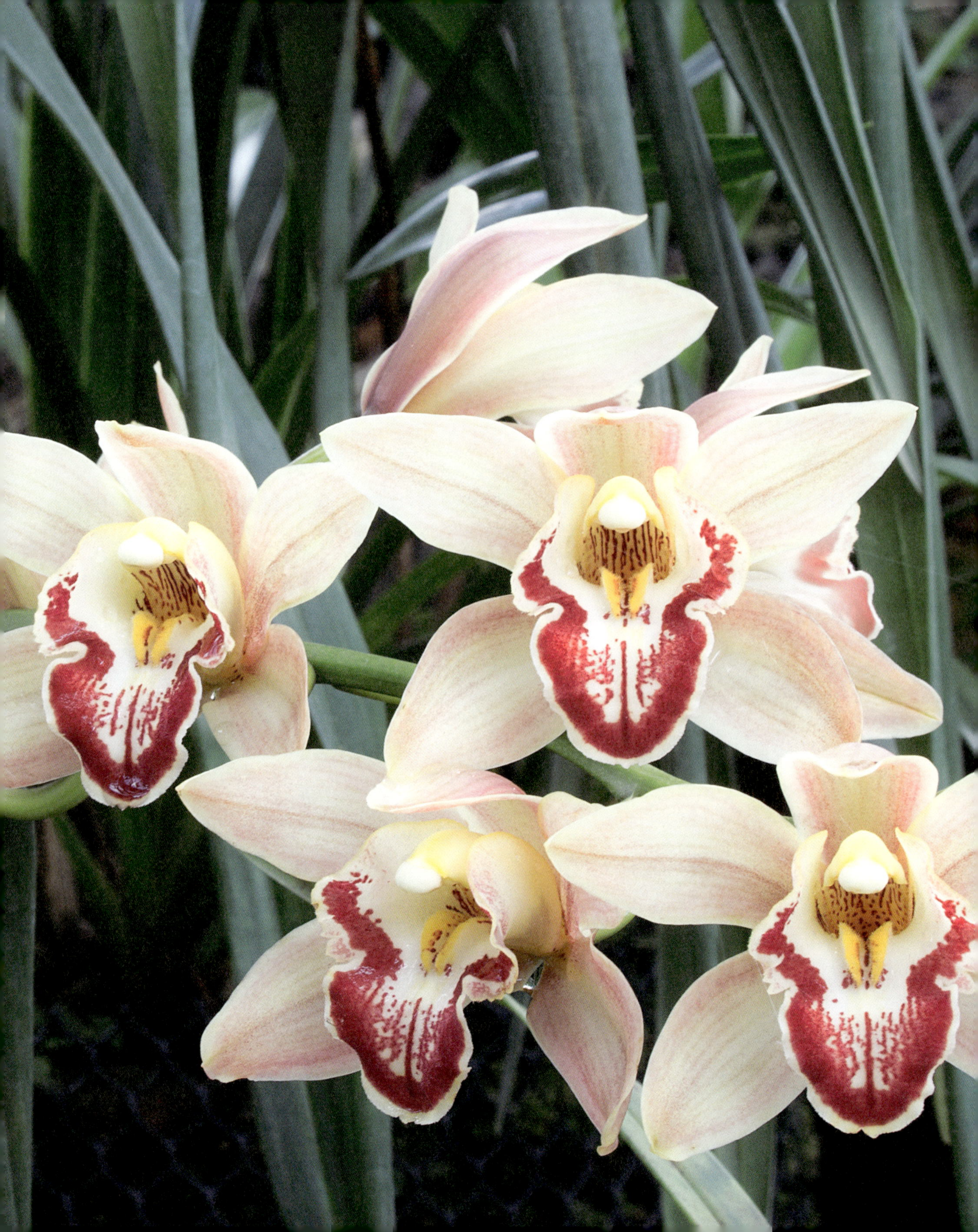

Cymbidium

BOAT ORCHID

BELOVED ORCHID FROM CHINA

Even Confucius was charmed by its fragrance and beauty. The Chinese have been cultivating *Cymbidium* since at least 500 BCE, considering it a symbol of prosperity, health, and longevity. The flower was also utilized in traditional Chinese medicine. Even today, *Cymbidium* remains one of the most popular orchids for cut flowers and decoration. Because the lip of the flower resembles a boat, the name *Cymbidium* comes from the Greek word *kymbos*, meaning "boat."

KAHNORCHIDEE

LIEBLINGSORCHIDEE AUS CHINA

Schon Konfuzius begeisterte sich für den Liebreiz dieser duftenden Orchidee. *Cymbidium* wurde in China seit mindestens 500 v. Chr. kultiviert. Sie galt als Symbol für Wohlstand, Gesundheit und ein langes Leben und fand in der Traditionellen Chinesischen Medizin Verwendung. Auch heute zählt *Cymbidium* zu den beliebtesten Schnitt- und Schmuckorchideen. Weil die Blütenlippen an ein Boot erinnern, nimmt ihr Trivialname ebenfalls Bezug auf das griechische Wort für „Kahn" – *kymbos*.

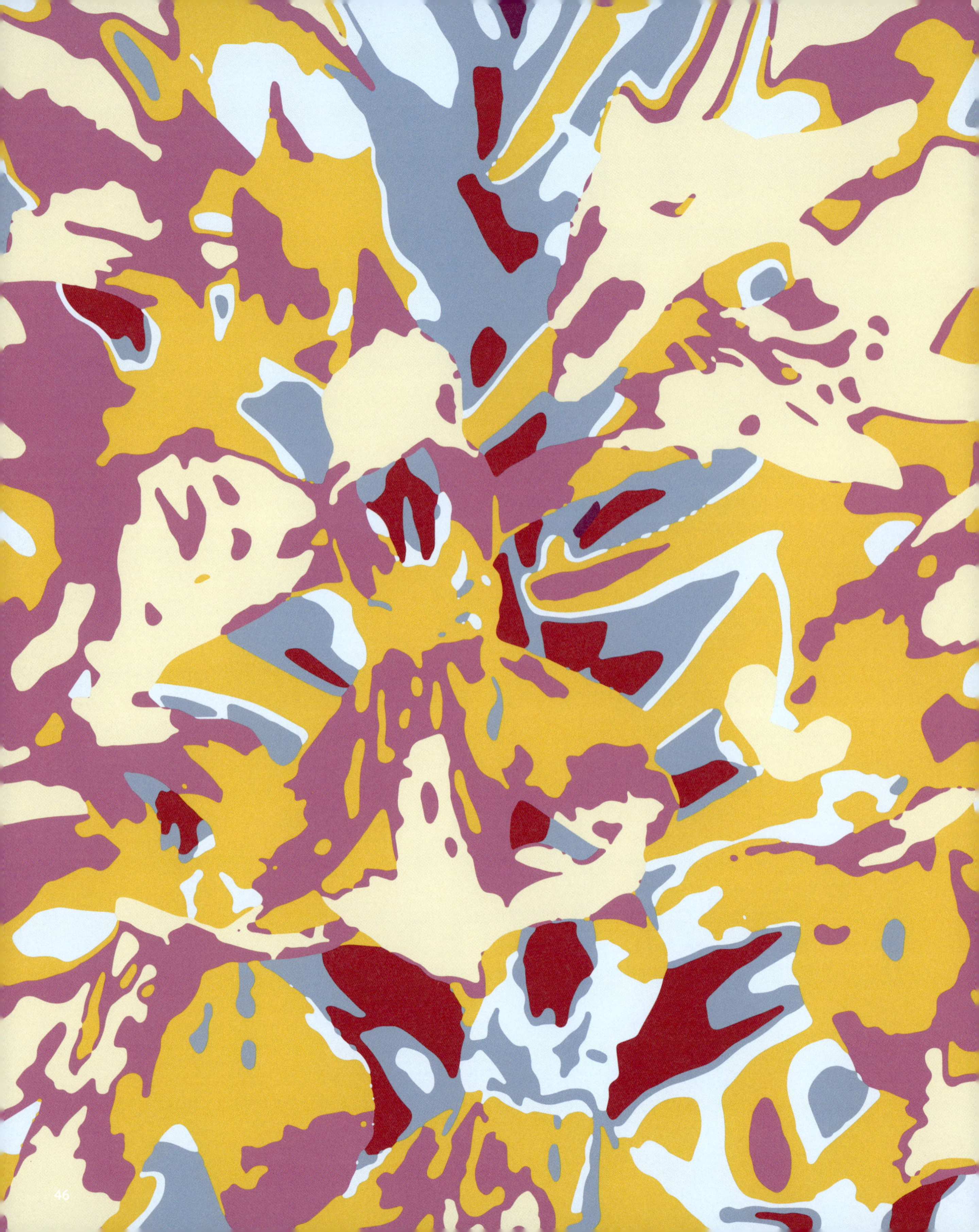

ORCHIDS IN THE NORTHERN HEMISPHERE

HEIMISCHE ORCHIDEEN

Untamed beauty worthy of protection

THE HAPPINESS OF FINDING AN ORCHID IN THE MEADOW

OF LADY'S SLIPPERS, RED HELLEBORINES, AND OTHER BEAUTIFUL SPECIES

People who think of orchids only as exotic beauties will be amazed at the variety of forms and color among the orchids of the temperate northern hemisphere. The lady's slipper orchid *(Cypripedium calceolus)*, with its unusual flower shape, looks more like it belongs in a tropical rainforest than in the beech forests of central Europe. The same goes for the filigree, colorful flowers of the red helleborine *(Cephalanthera rubra)*. In the United States, there are about 100 species of wild orchids; about half as many are found in the wild in the United Kingdom. Unfortunately, finding wild orchids is becoming increasingly difficult. For although wild orchids are generally protected, their existence is endangered. Some are threatened with extinction.

Unlike most tropical orchids, which are epiphytic or lithophytic (they grow on trees or rocks), wild orchids in most of North America and everywhere in Britain are terrestrial. (Some epiphytes are found in North America as far north as Florida.) The earth-dwelling orchids thrive in moist meadows and open forests. Unfortunately, these habitats are increasingly threatened by human activities such as intensive agriculture, with excessive use of fertilizers and pesticides. Additionally, rapid climate change is permanently altering orchid habitats. For example, what was once a well-watered meadow may dry out, making it impossible for a beautiful species like the broad-leaved marsh orchid *(Dactylorhiza majalis)* to survive there.

Wunderschön wild und schützenswert

DAS ORCHIDEENGLÜCK LIEGT IN DER WIESE

VON FRAUENSCHUHEN, WALDVÖGLEIN UND ANDEREN SCHÖNHEITEN

Wer Orchideen nur als exotische Schönheiten im Sinn hat, wird über die Farb- und Formenvielfalt einheimischer Orchideen erstaunt sein: Den Gelben Frauenschuh (*Cypripedium calceolus*) mit seiner außergewöhnlichen Blütenform würde man eher im tropischen Regenwald als in mitteleuropäischen Buchenwäldern vermuten, ebenso die filigranen, farbenprächtigen Blüten des Roten Waldvögleins (*Cephalanthera rubra*). In Deutschland gibt es etwa 70 verschiedene Orchideenarten, die draußen in der Natur zu finden sind. Soweit die Theorie – in der Praxis sieht es mit dem Auffinden wilder Orchideen nämlich immer schlechter aus: Obwohl wild wachsende Orchideen grundsätzlich unter Naturschutz stehen, ist ihr Bestand gefährdet und teils vom Aussterben bedroht.

Im Gegensatz zu den meist epiphytischen oder lithophytischen, also auf Bäumen oder Steinen wachsenden Tropenorchideen sind mitteleuropäische Orchideen terrestrisch: Die Erdbewohner lieben feuchte Wiesen und lichte Wälder. Doch diese Lebensräume gefährdet der Mensch immer mehr durch intensive Landwirtschaft mit Überdüngung und Pflanzenschutzmitteln. Auch der rapide fortschreitende Klimawandel verändert die Lebensbedingungen dauerhaft: Wenn ehemals feuchte Wiesen austrocknen, können Schönheiten wie das Breitblättrige Knabenkraut (*Dactylorhiza majalis*) dort nicht mehr überleben.

NEXT PAGE | *Cephalanthera rubra* (left) and *Dactylorhiza majalis* (right)
NÄCHSTE SEITE | Rotes Waldvöglein (links) und Breitblättriges Knabenkraut (rechts)

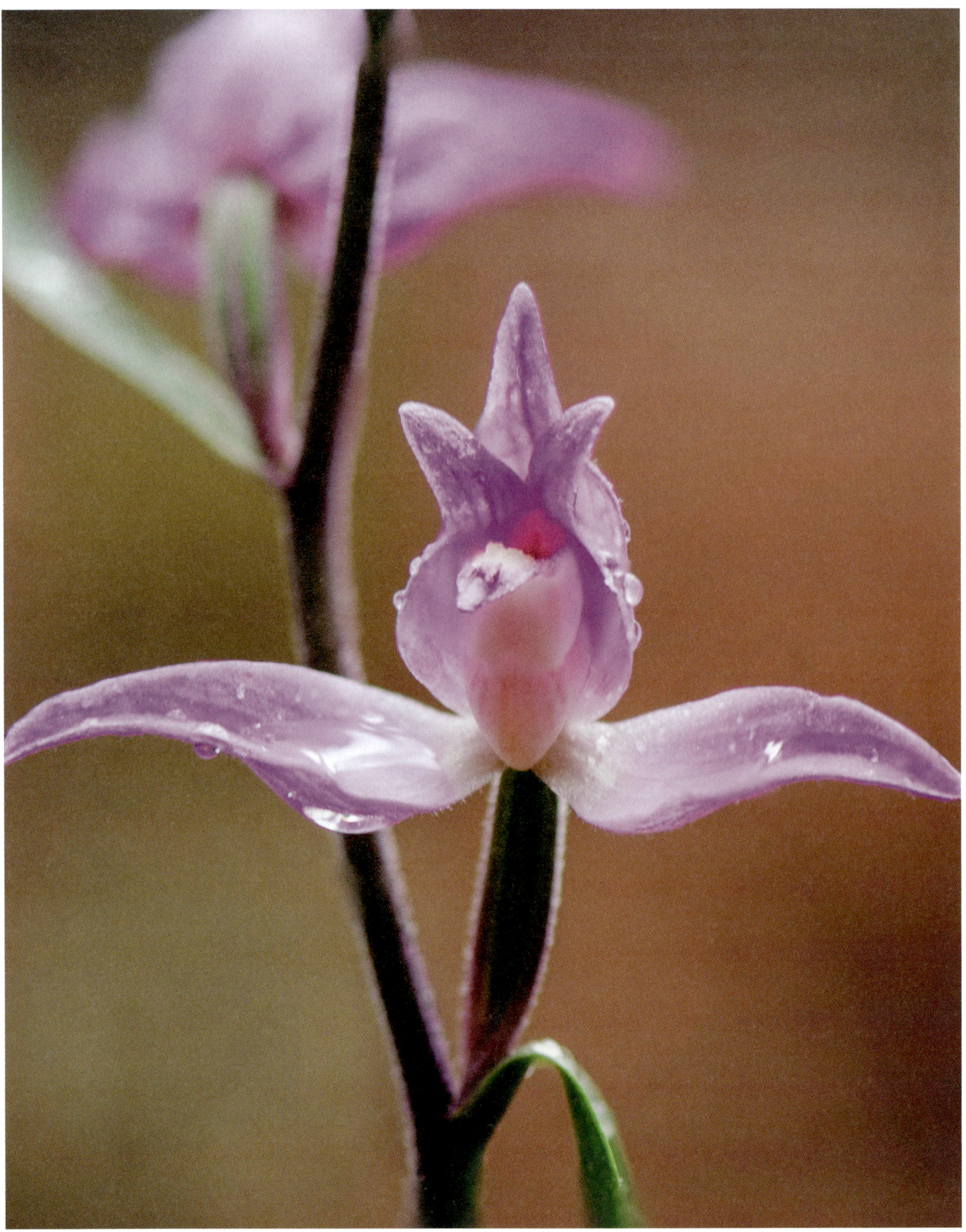

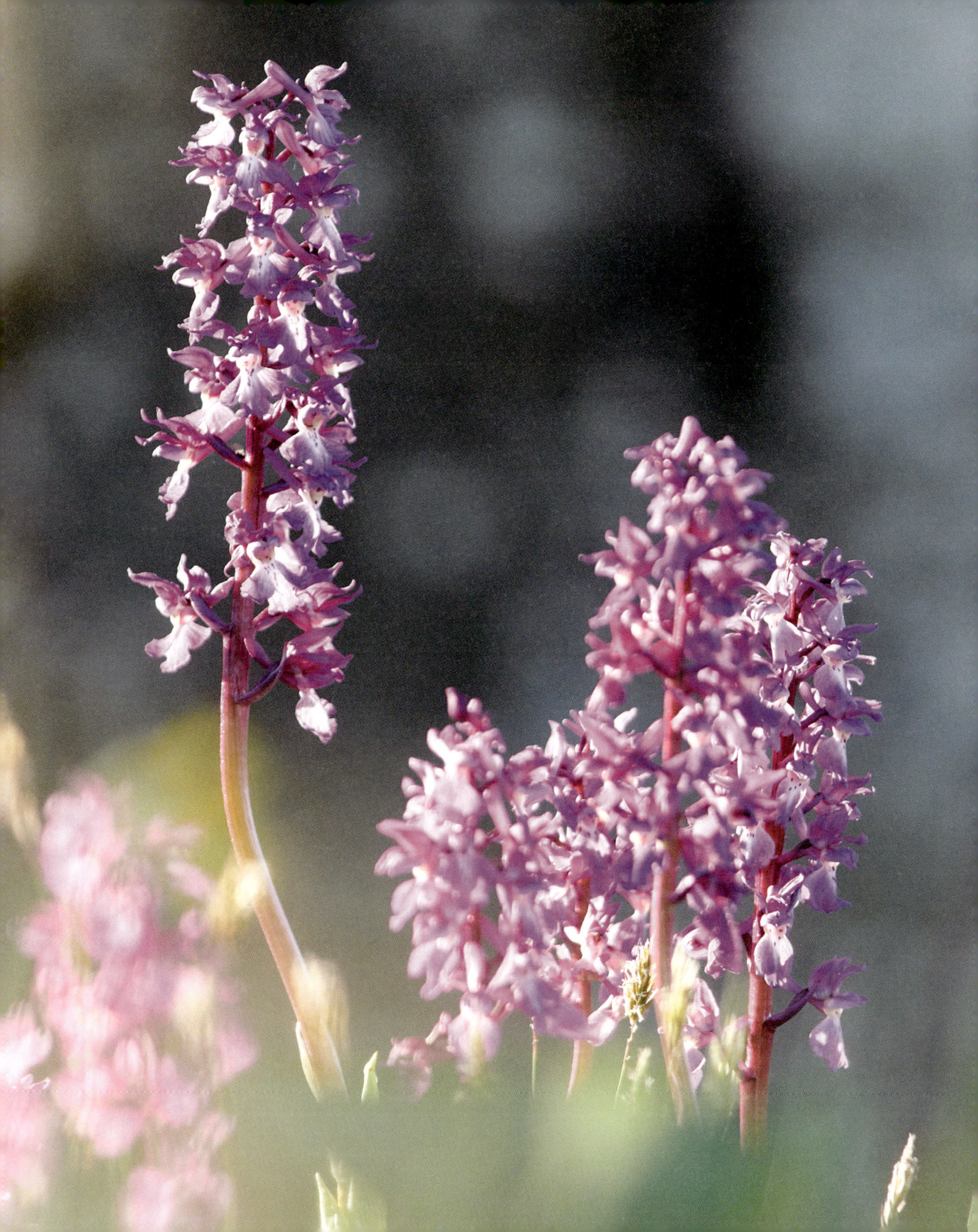

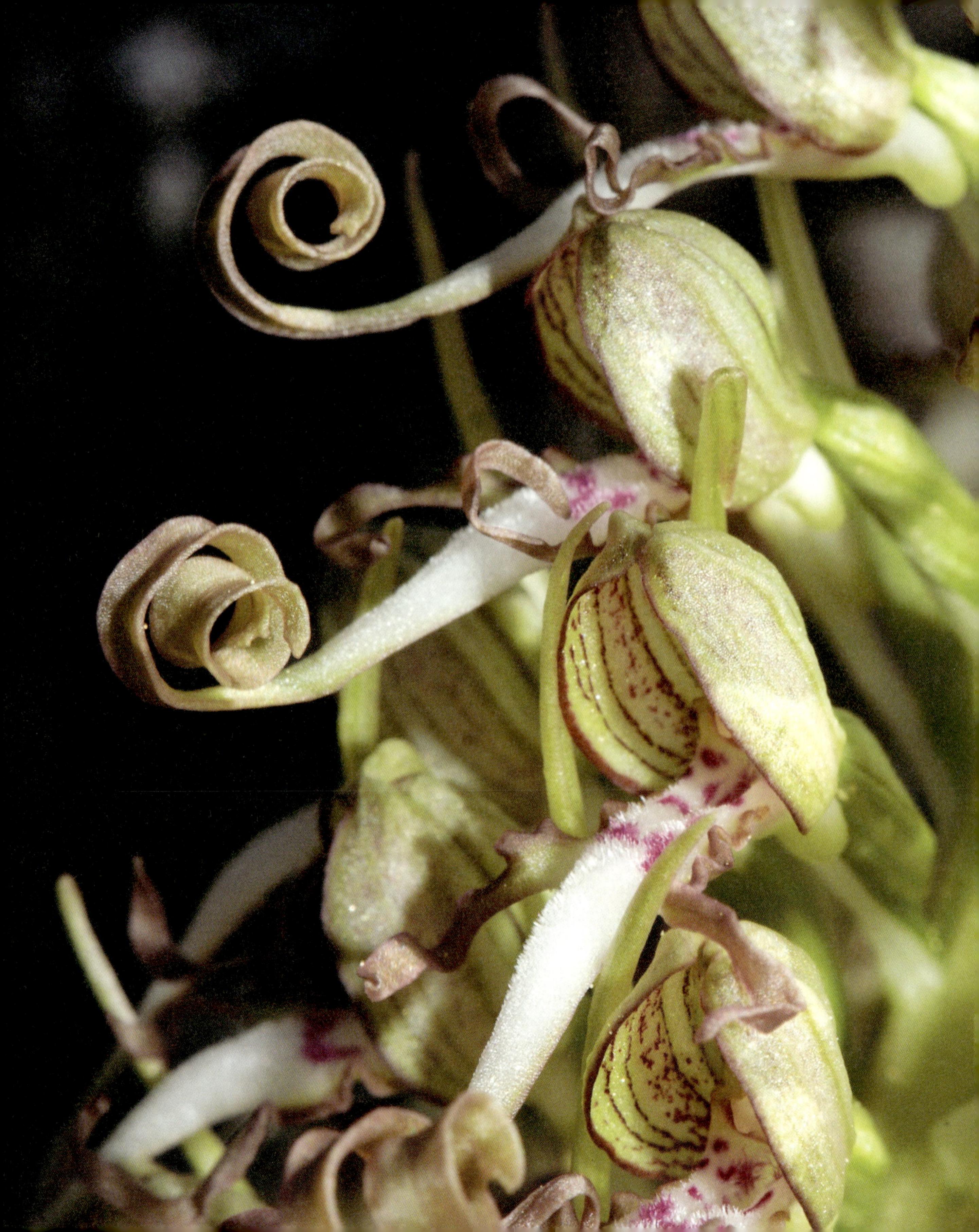

Anything but Latin

POPULAR NAMES

FANTASTIC BEASTS AND WHERE TO FIND THEM

Goat orchid, monkey orchid, bee orchid, butterfly orchid, frog orchid, lizard orchid *(left page)* or even leopard orchid. At least where popular names are concerned, there is an intriguing tendency toward zoomorphism, the inclination to engage in animal symbolism. Generally the plant's appearance furnishes an immediate explanation for the name; sometimes, the likeness is so obvious that an insect, "aped" by the orchid, will confuse the flower for a member of its own species. Luckily, the popular names also sometimes refer to the plant's preferred habitat, making them a bit easier to spot: bog orchid, fen orchid, or the even more specific Lindisfarne orchid.

Alles außer Latein

ORCHIDEEN IM VOLKSMUND

KURIOSE KOMBINATIONEN UND „SPRECHENDE" NAMEN

Deutsche Namen heimischer Orchideen lassen uns oft schmunzeln. Bei den kuriosen Kombinationen denkt man gern an urige Schimpfwörter: Hummel-Ragwurz, Übersehene Fingerwurz, Bocks-Riemenzunge *(linke Seite)* ... Neben Referenzen auf das Erscheinungsbild weisen Namensbestandteile auch oft auf den Standort hin (z. B. Sumpf-Knabenkraut) oder verraten die Blütezeit: Die Schnepfen-Ragwurz blüht im Frühling zur Brutzeit der Schnepfe, das auch Kuckucksblume genannte Breitblättrige Knabenkraut im Mai, wenn der Kuckuck singt.

Cypripedium calceolus

LADY'S SLIPPER ORCHID

BRIGHT STAR AMONG THE ORCHIDS OF CENTRAL EUROPE

The yellow lady's slipper is an absolute stunner, one of the most amazing orchids you'll ever see in central Europe. The *Cypripedium calceolus* boasts a blend of bright lemon-yellow and purplish hues, and it flowers only briefly. In southern Germany, for example, it only blooms from May to June, and some people in those parts refer to it as the Whitsun flower or Mary's beauty. Its flowers, which can grow up to eight centimeters in size, make this orchid the belle of the ball in Europe. And it's quite the diva, a true creature of the limelight, that prefers bright locations and is no fan of dimly-lit environs or acidic soil. It likes heights and is comfortable even at over 2,000 m (more than 6,500 feet) of elevation. And it feels right at home in Scandinavia.

This orchid is called the *Cypripedium* due to the bulbous, bright yellow lip blossom that resembles a lady's slipper. This is actually a kettle-shaped trap that lures insects with an intense fragrance, enticing them to move from slipper to slipper, pollinating the flowers as they crawl out. The sweet scent is a mean trick, though, as the flower fails to deliver. There is no nectar for the bees, which are merely supposed visit as many plants as possible as the orchid cannot be fertilized through self-pollination.

Cypripedium calceolus

GELBER FRAUENSCHUH

DER STAR DER MITTELEUROPÄISCHEN ORCHIDEEN

Als eine der prächtigsten Wildorchideen Mitteleuropas gilt der Gelbe Frauenschuh: In einem Farbmix aus Zitronengelb und Purpur blüht *Cypripedium calceolus* z. B. im Süden Deutschlands nur kurz von Mai bis Juni und wird daher dort mancherorts Pfingstblume oder Marienschön genannt. Die bis zu acht Zentimeter großen Blüten machen den Gelben Frauenschuh zum Star der europäischen Orchideen. Eine kleine Diva ist sie schon: Sie liebt das (Rampen-)Licht und mag es weder dunkel noch sauer, will dafür aber hoch hinaus – sogar in über 2000 Metern Höhe und in Skandinavien fühlt sie sich wohl.

Ihren Namen bezieht die *Cypripedium* aus der leuchtend gelben, wie ein Damenpantoffel geformten bauchigen Lippenblüte. Diese dient als „Kesselfalle": Durch den intensiven Blütenduft werden Insekten von Pantoffel zu Pantoffel gelockt und bestäuben beim Herauskrabbeln automatisch die Blüte. Ganz schön fies: Trotz des süßen Dufts wartet noch nicht einmal Nektar auf die Insekten – so sollen die Bienen hungrig möglichst viele Pflanzen besuchen, da die Orchidee nicht über Selbstbestäubung befruchtet werden kann.

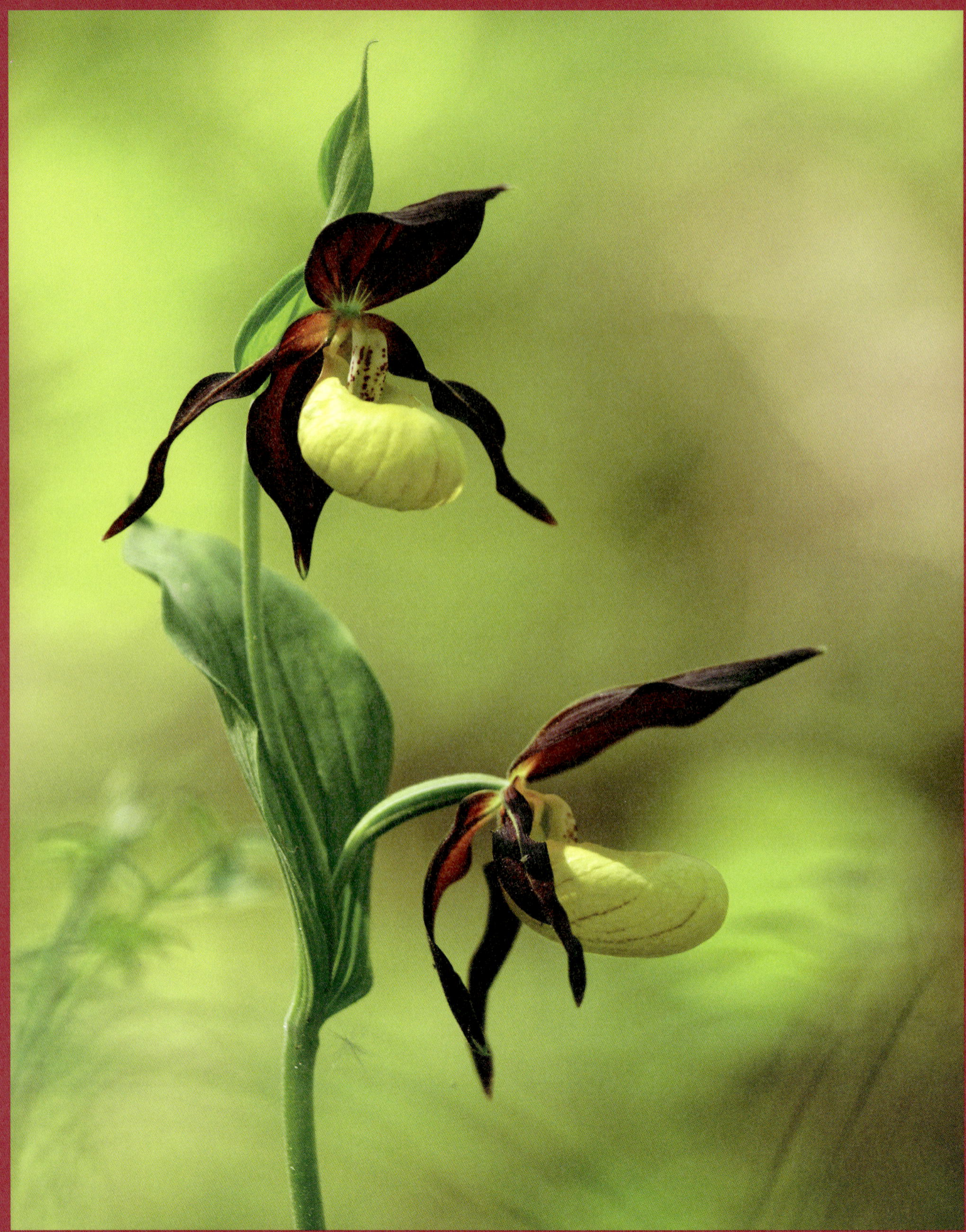

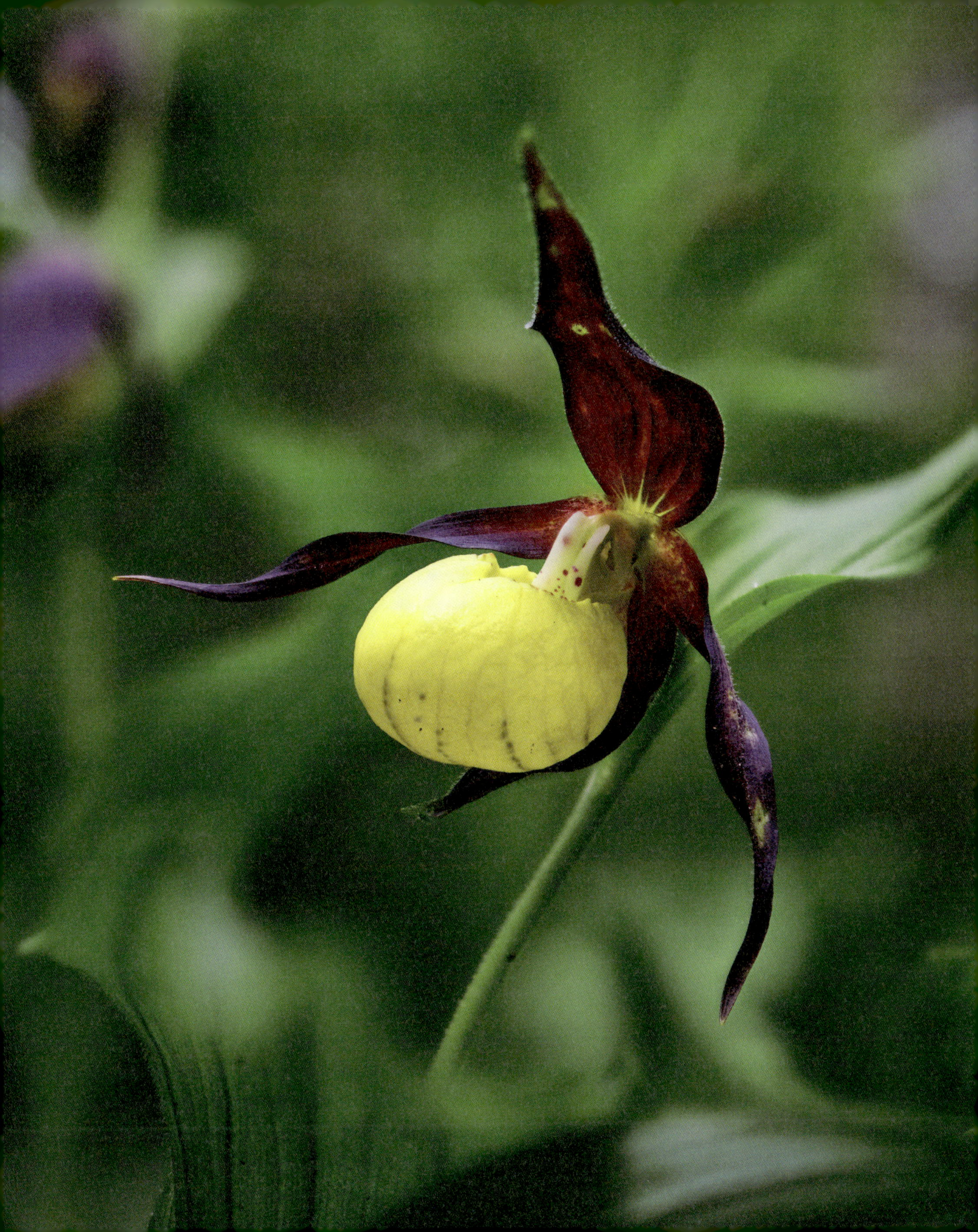

The name says it all

ORCHIS

THE TUBERS OF THIS ORCHID ARE EVEN MORE COVETED THAN THE STUNNING FLOWERS

These various orchids are widespread in Europe, but they're also some of the most endangered species out there. Many Orchis species – the pale orchid, the purple orchid, the military orchid – are under pressure because of unfavorable conditions in their habitats. For example, many forest understories are becoming overgrown, which prevents enough light from reaching terrestrial orchids. To make matters worse, the dense undergrowth attracts wild boars that dig up the tasty roots. Droughts in winter as well as increasingly acidic soils due to overfertilization are also taking a toll on the nutrient balance of wild orchids, making some of them gravely endangered species.

Sadly, and as magnificent as they are, Orchis and company are also popular with plant poachers, who not only pick but may also completely uproot them. That orchids are coveted is not new, though; their tuberous roots were once considered aphrodisiac, which explains, for example, the German regional name Liebeswurz ("lovewort"). Pregnant women used to consume orchids because of their resemblance to testicles (*orchis* in Greek). This was believed to ensure the birth of a male heir.

Orchis

KNABENKRAUT

NOCH BEGEHRTER ALS DIE WUNDERSCHÖNEN BLÜTEN SIND DIE KNOLLEN DIESER ORCHIDEE

Knabenkraut-Orchideen sind in Europa geografisch weit verbreitet und gehören doch zu den gefährdetsten Arten. Ob Blasses Knabenkraut, Purpur-Knabenkraut oder Helm-Knabenkraut – die Biotope vieler *Orchis*-Arten gerieten in den letzten Jahrzehnten durch ungünstige Einflüsse in Bedrängnis: Viele Wälder verbuschen, sodass nicht mehr genug Licht zu den Erdorchideen durchdringt – stattdessen aber Wildschweine, die die schmackhaften Knollen ausbuddeln. Trockenheit im Winter und durch Überdüngung immer saurer werdende Böden belasten zusätzlich den Nährstoffhaushalt der Wildorchideen und schicken sie zunehmend auf die Rote Liste der gefährdeten Arten.

Zu allem Überfluss sind die prachtvoll blühenden Knabenkraut-Orchideen begehrt bei Pflanzenräubern, die sie nicht nur pflücken, sondern komplett ausgraben. Objekt der Begierde waren *Orchis* schon in vergangenen Zeiten: Ihre Knollen galten als Aphrodisiakum, weswegen die Knabenkraut-Orchideen mancherorts auch heute noch Liebeswurz genannt werden. Aufgrund ihrer namensgebenden Ähnlichkeit zu männlichen Hoden (auf griechisch *orchis*) war ihr Verzehr bei Schwangeren beliebt: So sollte für die Geburt eines männlichen Stammhalters gesorgt werden.

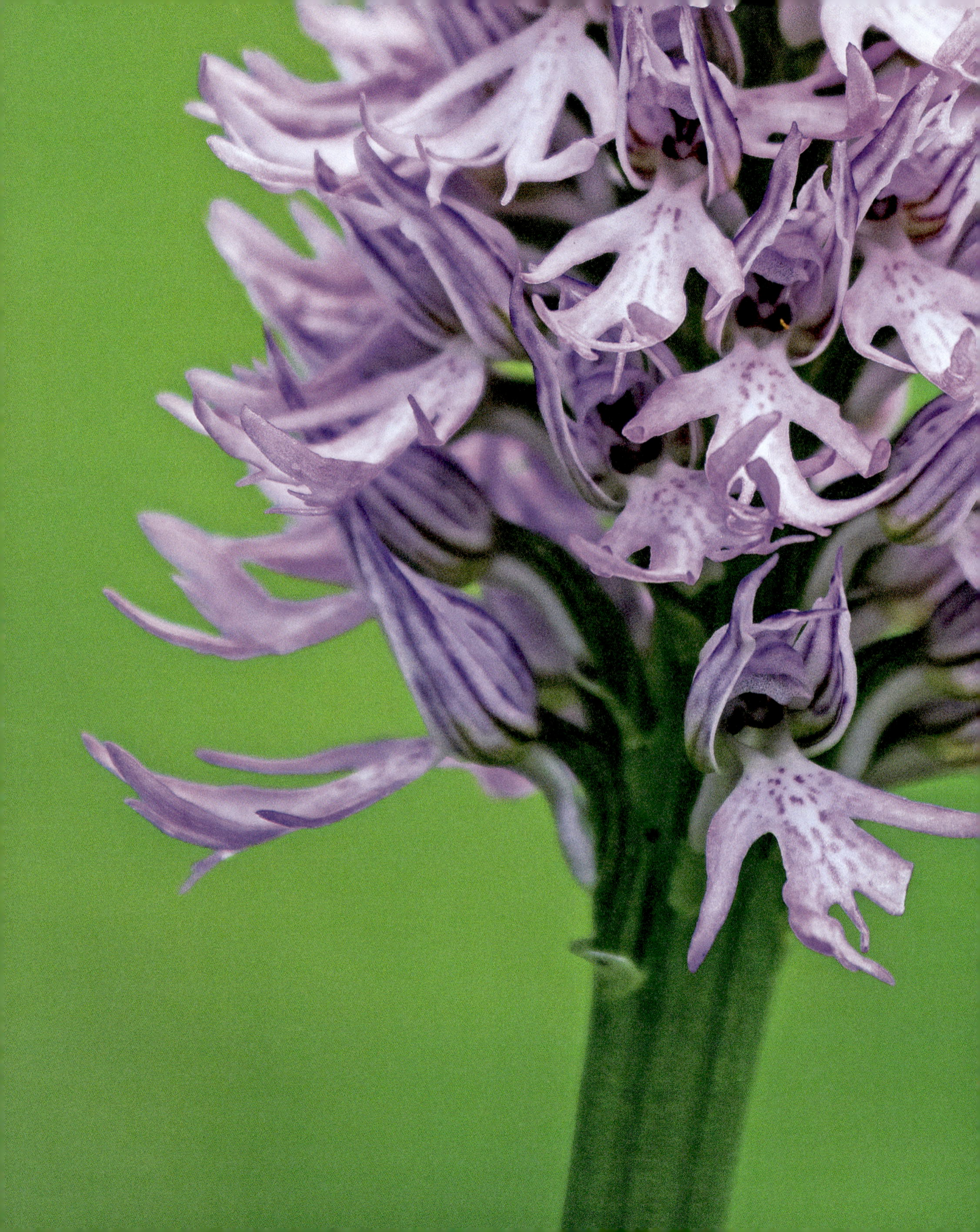

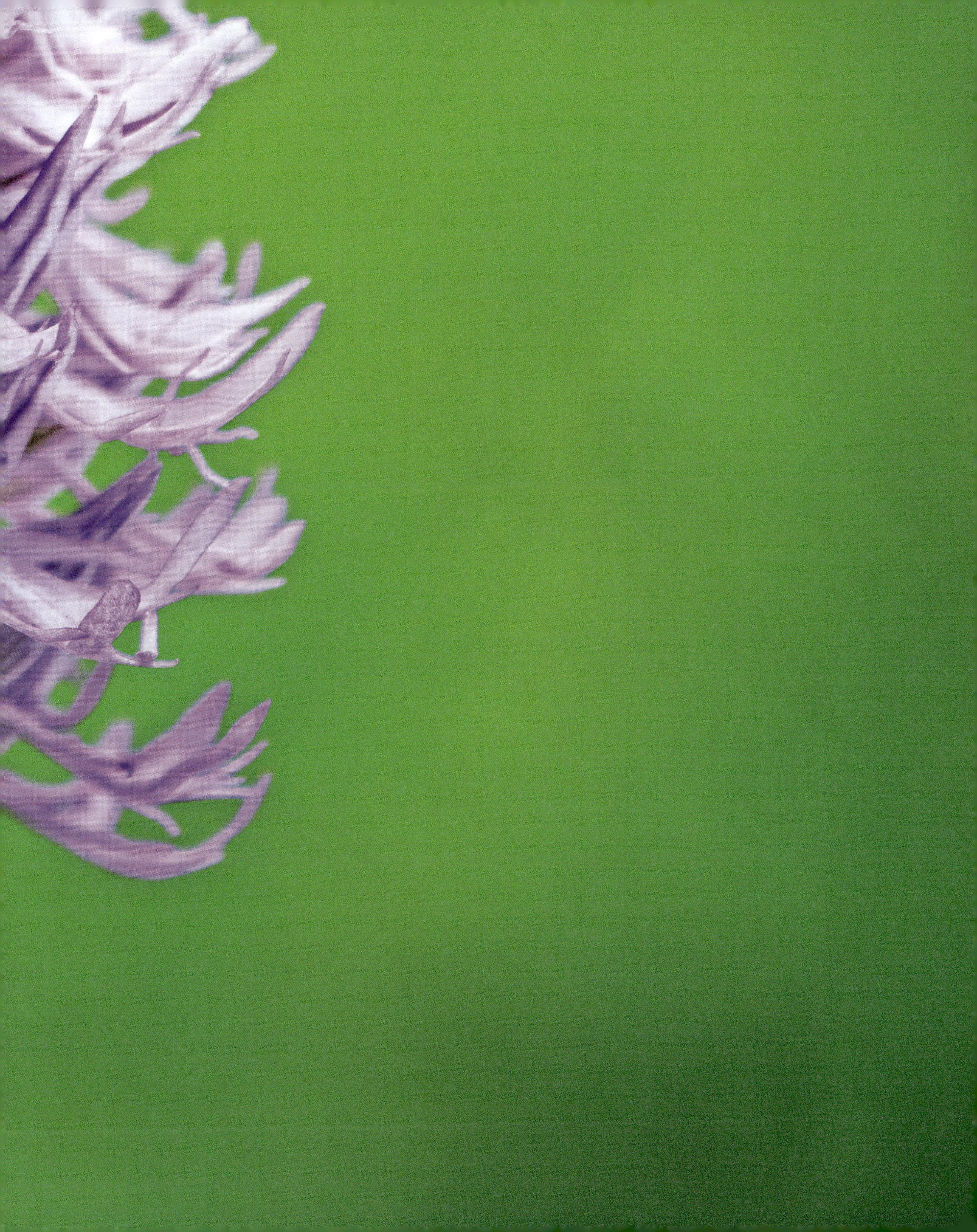

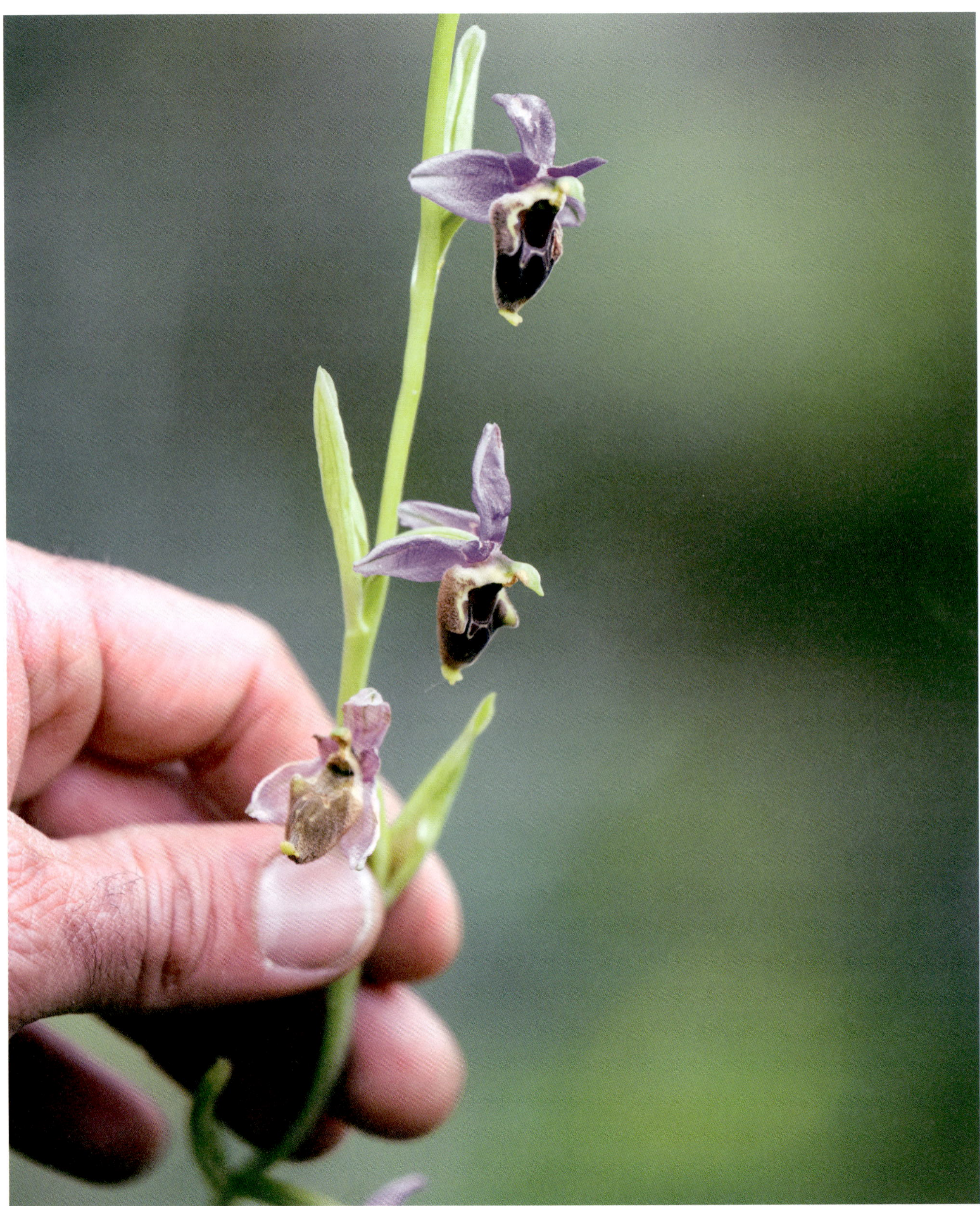

Ophrys

BEE ORCHID

AN INGENIOUS SCHEME

It's the furry edges of the labellum that inspired both the botanical and the popular name for this beauty. Whereas *Ophrys* points to the Greek word for "eyebrow", the bee orchid takes its common name from the resemblance to furry insect bodies. But bee orchids are full of tricks when it comes to their reproductive strategy: the late spider orchid, the sawfly orchid, and the fly orchid are all consummate mimics, imitating not only the appearance but also the pheromones of the eponymous female insects and thus attracting the males of these species for pollination.

RAGWURZ

GANZ SCHÖN RAFFINIERT

Aus der Knolle, der „Wurz" der *Ophrys*, ragt der Blütenstängel nach oben – soweit die recht einfache Erklärung des Volksnamens Ragwurz. Raffiniert hingegen ist ihre trickreiche Fortpflanzungsstrategie: Hummel-Ragwurz, Wespen-Ragwurz und Fliegen-Ragwurz sind Meister der Mimikry! Sie imitieren nicht nur das Aussehen, sondern auch die Sexualduftstoffe der namensgebenden weiblichen Insekten und locken so die Männchen dieser Arten zur Bestäubung an.

Cephalanthera

HELLEBORINES

STUNNINGLY DELICATE

The graceful form of *Cephalanthera* looks wispy and wing-like, which is why some North American species are poetically called "phantom" orchids. With up to twenty-five individual flowers blooming together on a long-leaved helleborine, it looks as though a flock of tiny birds has just settled on the stems. The red helleborine orchid is extremely rare in Britain, and even where found, it may go for years without producing a bloom. But if the soil is alkaline enough, its purple flowers will shine even brighter.

WALDVÖGLEIN

ZAUBERHAFT ZART

Die elegante Silhouette der *Cephalanthera* erinnert an zarte Vogelflügel und inspirierte zum poetischen Namen Waldvöglein. In der Tat, wenn die bis zu 25 Einzelblüten des Langblättrigen Waldvögleins in voller Blüte stehen, wirkt es, als hätte sich gerade ein winziger Vogelschwarm auf den Stängeln niedergelassen. Eine rare Schönheit ist das Rote Waldvöglein: Bisweilen blüht eine Pflanze mehrere Jahre nicht. Ist der Boden aber reich an Kalk, leuchten ihre purpurroten Blüten umso kräftiger.

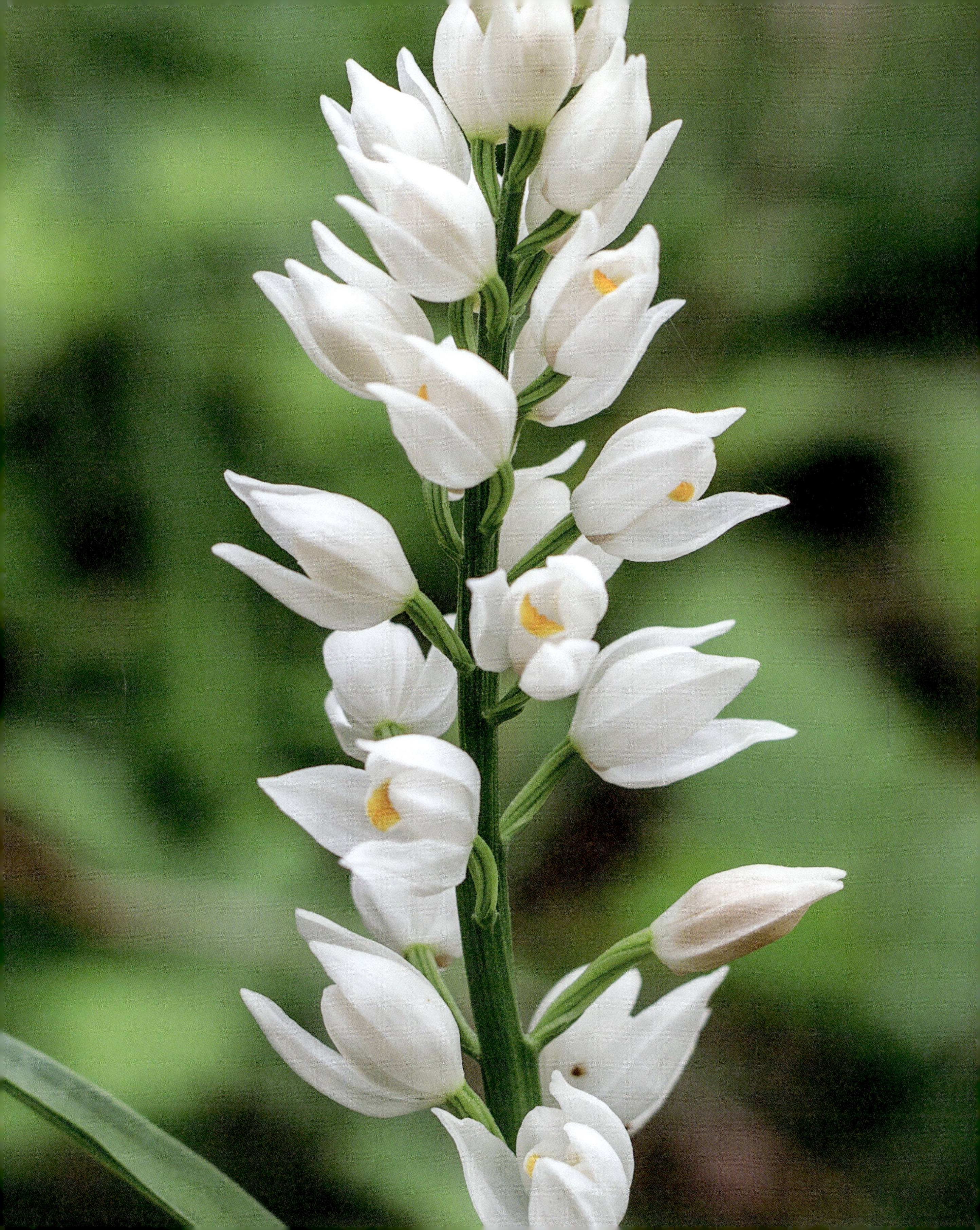

Neottia nidus-avis

BIRD'S NEST ORCHID

THE DIAMOND IN THE ROUGH

The flowers and stems of *Neottia nidus-avis* are a pale beige, with no green leaves in sight. How does this orchid survive when it doesn't photosynthesize? It's thanks to a special partnership with a soil fungus that completely supplies it with the nutrients it needs. The fungus, in turn, obtains these nutrients from surrounding tree roots. *Neottia nidus-avis* is named after its strong root rhizome, whose strands are intertwined like a bird's nest.

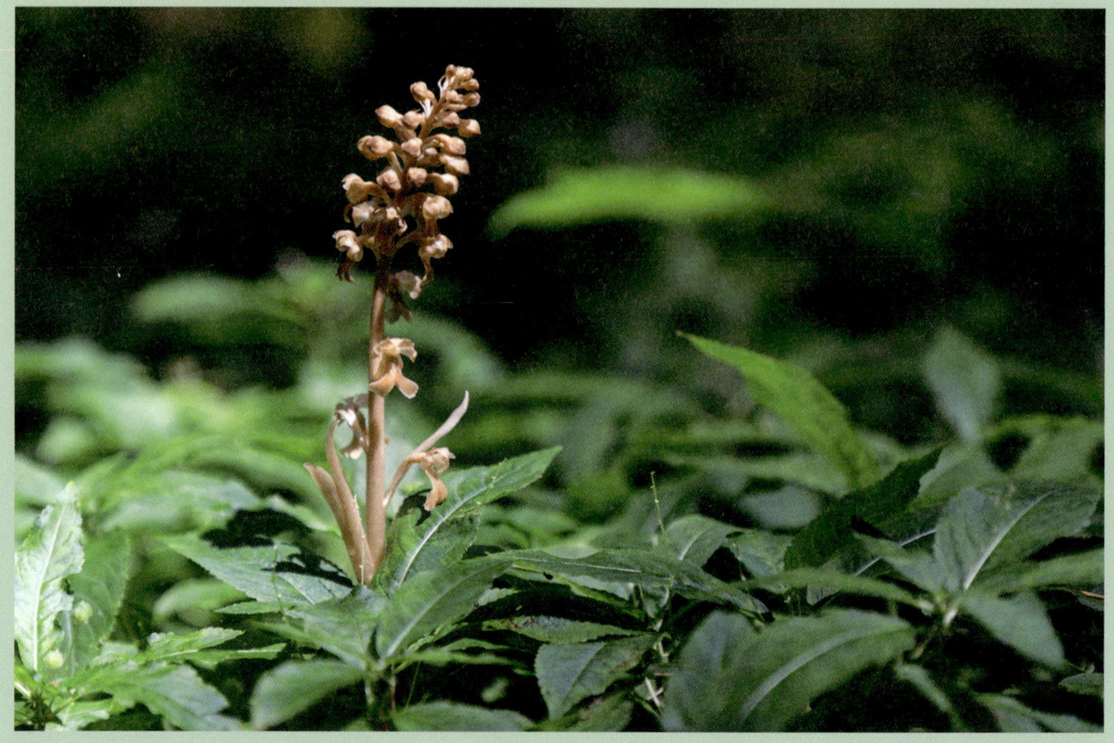

Neottia nidus-avis

VOGELNESTWURZ

HIER TRIFFT EINE REDENSART PERFEKT: NICHT SCHÖN, ABER SELTEN

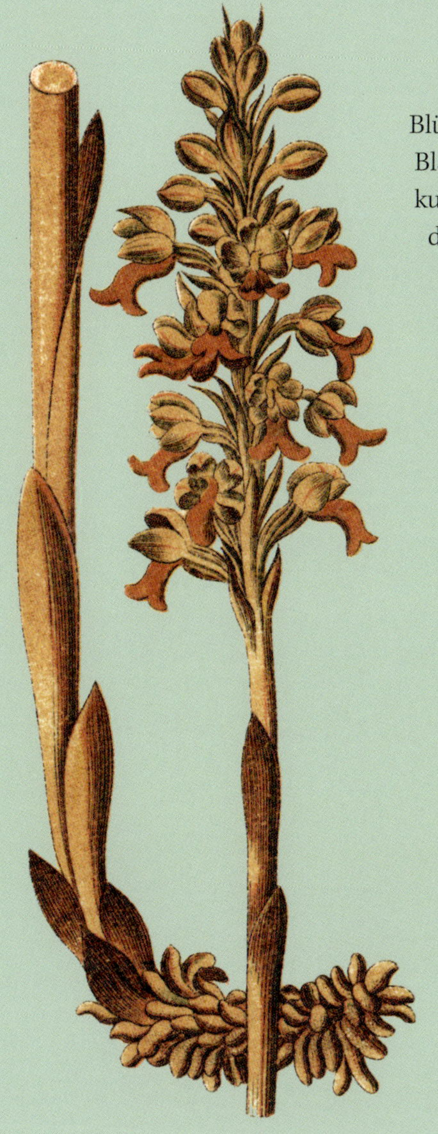

Blüten und Stängel in fahlem Beige, weit und breit kein Blattgrün – wie kann die *Neottia nidus-avis*, die so offenkundig keine Photosynthese betreibt, überleben? Nur dank einer ganz speziellen Partnerschaft: Ein Bodenpilz umgibt ihr Rhizom und versorgt sie komplett mit Nährstoffen. Die wiederum holt sich der Pilz aus umliegenden Baumwurzeln. Namensgeber für *Neottia nidus-avis*, zu Deutsch Vogelnestwurz, ist übrigens ihr kräftiges Wurzelrhizom, dessen Stränge wie ein Nest dicht ineinander verschlungen sind.

Anacamptis

PYRAMIDAL ORCHID

FAMOUS FOR ITS BEAUTIFUL PYRAMIDS OF FLOWERS

The botanical name *Anacamptis* was given to them, in 1817, because their bracts bend back under the individual flowers (*anakamptein* in Greek). Keep an eye out for them while hiking, as they can grow in all sorts of challenging locations, from shorelines with winds and high moisture to alpine settings of up to 2,300 meters (7,500 feet) in elevation. Although they are survivors practically everywhere, they don't grow very large in such punishing environments.

HUNDSWURZ

WUNDERSCHÖNE BLÜTENPYRAMIDEN

Weil sich die Tragblätter dieser Orchidee unter den einzelnen Blüten zurückbiegen (auf Griechisch *anakamptein*), erhielt sie 1817 den botanischen Namen *Anacamptis*. Die deutschen Namen der *Anacamptis*-Orchideen sorgen für etwas Verwirrung: Einige ihrer Vertreter werden als Knabenkraut bezeichnet, da diese erst 1997 anhand ihrer genetischen Merkmale den *Anacamptis* zugeordnet wurden. Als Pyramiden-Hundswurz oder Spitzorchis ist seit jeher die leuchtend purpurrote *Anacamptis pyramidalis* bekannt.

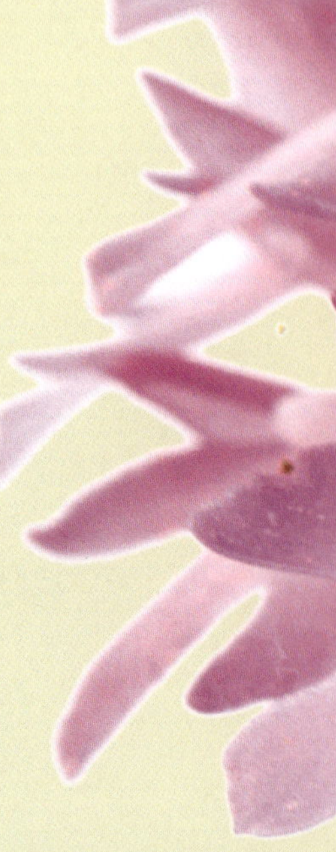

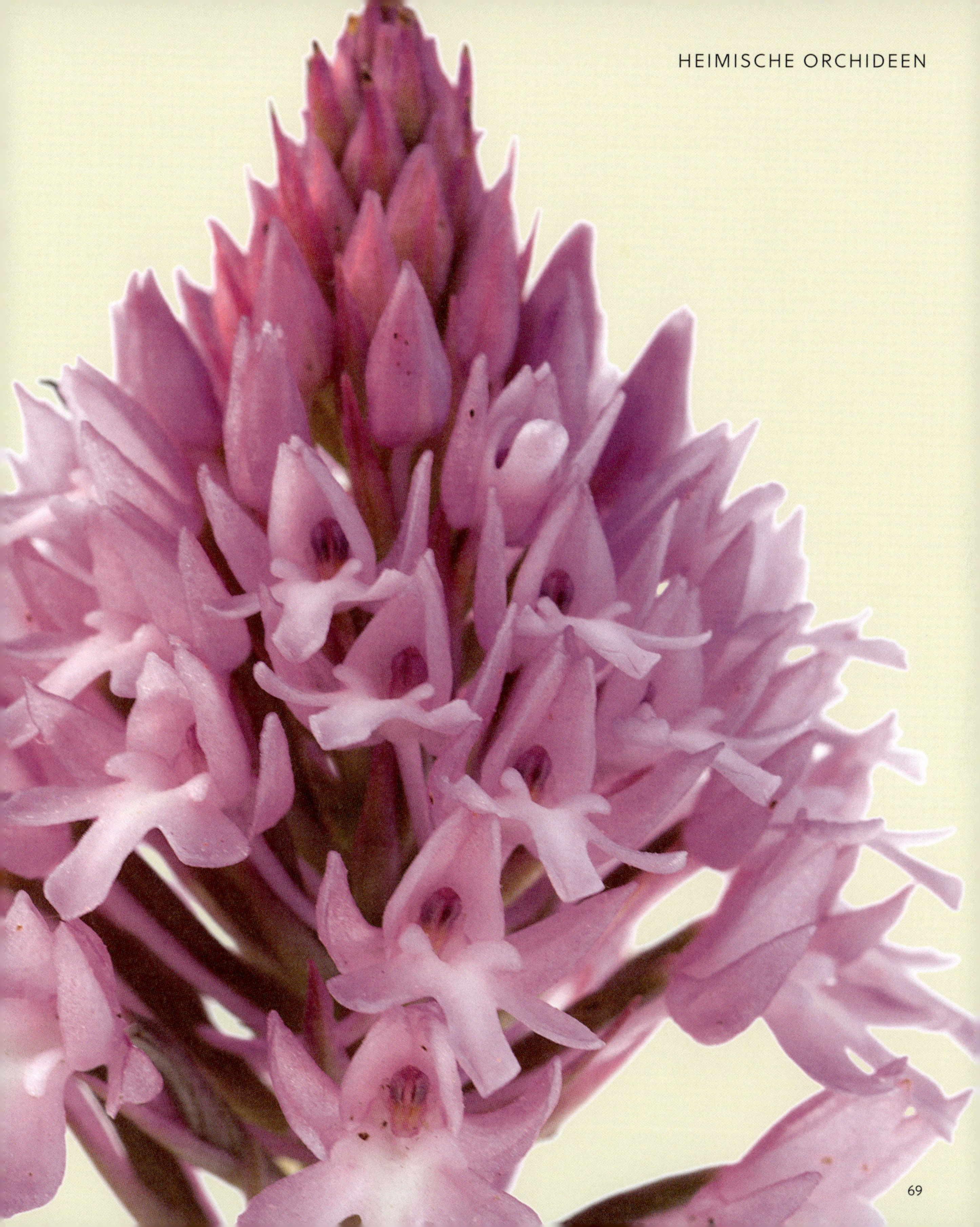

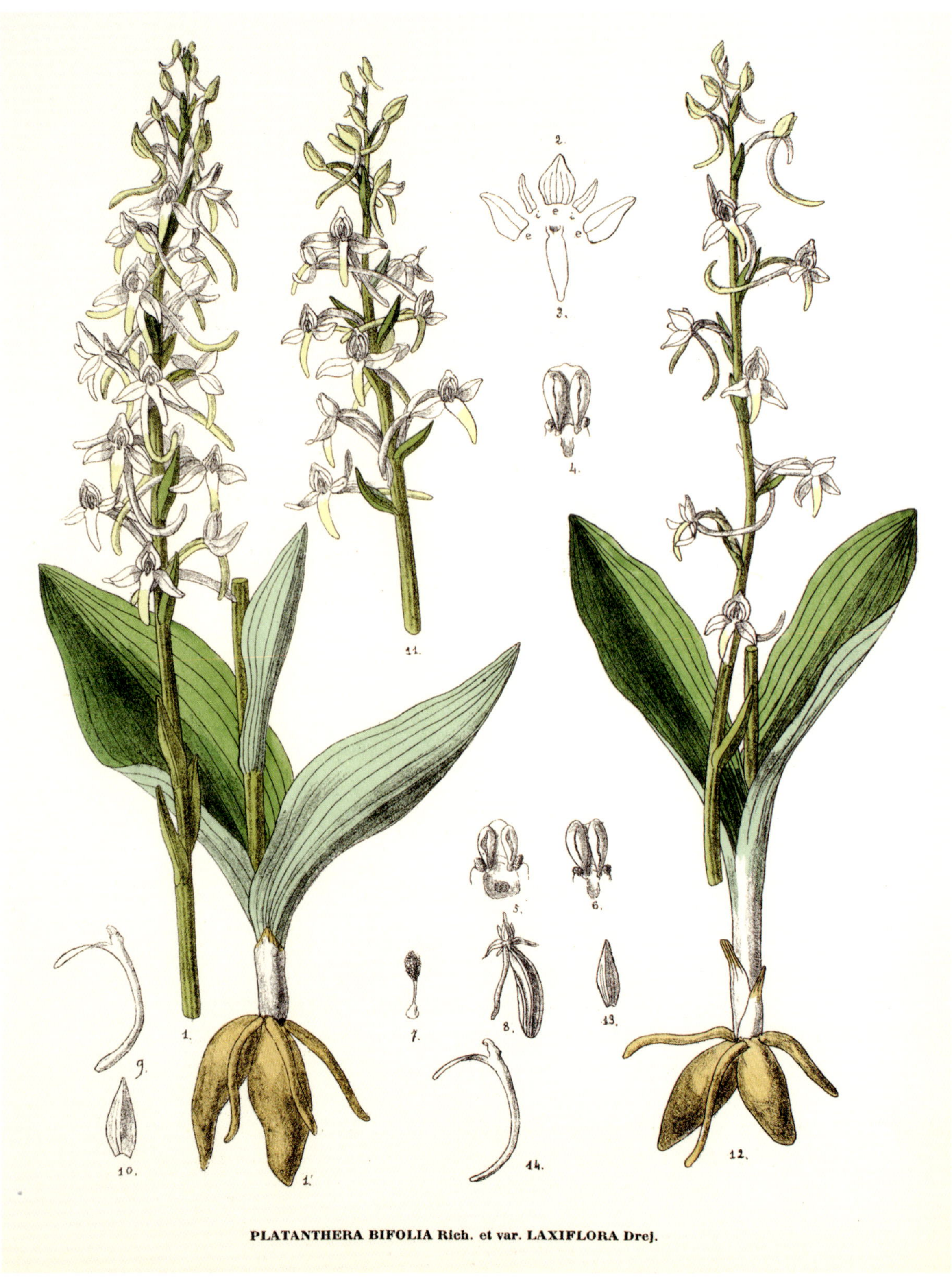

PLATANTHERA BIFOLIA Rich. et var. LAXIFLORA Drej.

Platanthera

BUTTERFLY ORCHID

DAMSELS IN THE MAYTIDE

As the orchid spreads its petals and sepals like the wings of a butterfly, it's easy to see how the popular names, of "lesser" and "greater butterfly" orchid, came about for these two of the most common types of *Platanthera*. The names of orchids can be confusing for the not-overly-zoologically-minded, who may tend to mix up moths and butterflies. But no, butterfly orchids are not related to *Phalaenopsis (see page 22)*, which are commonly known as moth orchids. However, *Platanthera bifolia*, the lesser butterfly orchid, is very attractive for nocturnal moths as it only exudes its scent at night.

WALDHYAZINTHEN

SCHÖNE MAID DER MAIENZEIT

Auch bei den Waldhyazinthen gibt der Name Anlass zur Verwirrung, täuscht hier doch das Aussehen der traubenartig angeordneten Blüten der *Platanthera* eine Beziehung zur Gattung der Hyazinthen vor. Weil sie im Mai, wenn der Kuckuck zu singen beginnt, erblüht, kennt man sie auch als Kuckucksblume. Neben der Grünen Waldhyazinthe kommt in mitteleuropäischen Bergregionen noch die Weiße Waldhyazinthe vor: *Platanthera bifolia* verströmt ihren Duft nur nachts und zieht dadurch auch nur nachtaktive Schmetterlinge an.

Himantoglossum hircinum

LIZARD ORCHID

QUITE CHEEKY

Among the Mediterranean, heat-loving *Himantoglossum* orchids, the lizard orchid is native to central Europe. Its Greek-derived name, from *himas* for "strap" plus *glossa* for "tongue", is readily explained: in full bloom, the flower's long, unfurled central labellum looks like a tongue sticking out at the observer. But anyone who knows that *hircus* is Latin for "goat" may quite rightly wonder how alluring the scent of this orchid could possibly be, and as a matter of fact, *Himantoglossum hircinum* gives off an intense goaty odor.

RIEMENZUNGE

GANZ SCHÖN FRECH

Unter den mediterranen, wärmeliebenden *Himantoglossum* ist vor allem die Bocks-Riemenzunge in Mitteleuropa heimisch. Deren Namensgebung (*himas* für Riemen und *glossa* für Zunge kommen aus dem Griechischen) ist leicht erklärt: In voller Blüte rollt sich der mittlere Blütenlappen lang aus und streckt dem Betrachter „die Zunge aus". Wer weiß, dass *hircus* lateinisch für Ziegenbock steht, fragt sich zu Recht, ob der Duft dieser Orchidee betörend ist: Tatsächlich riecht *Himantoglossum hircinum* intensiv nach Ziege!

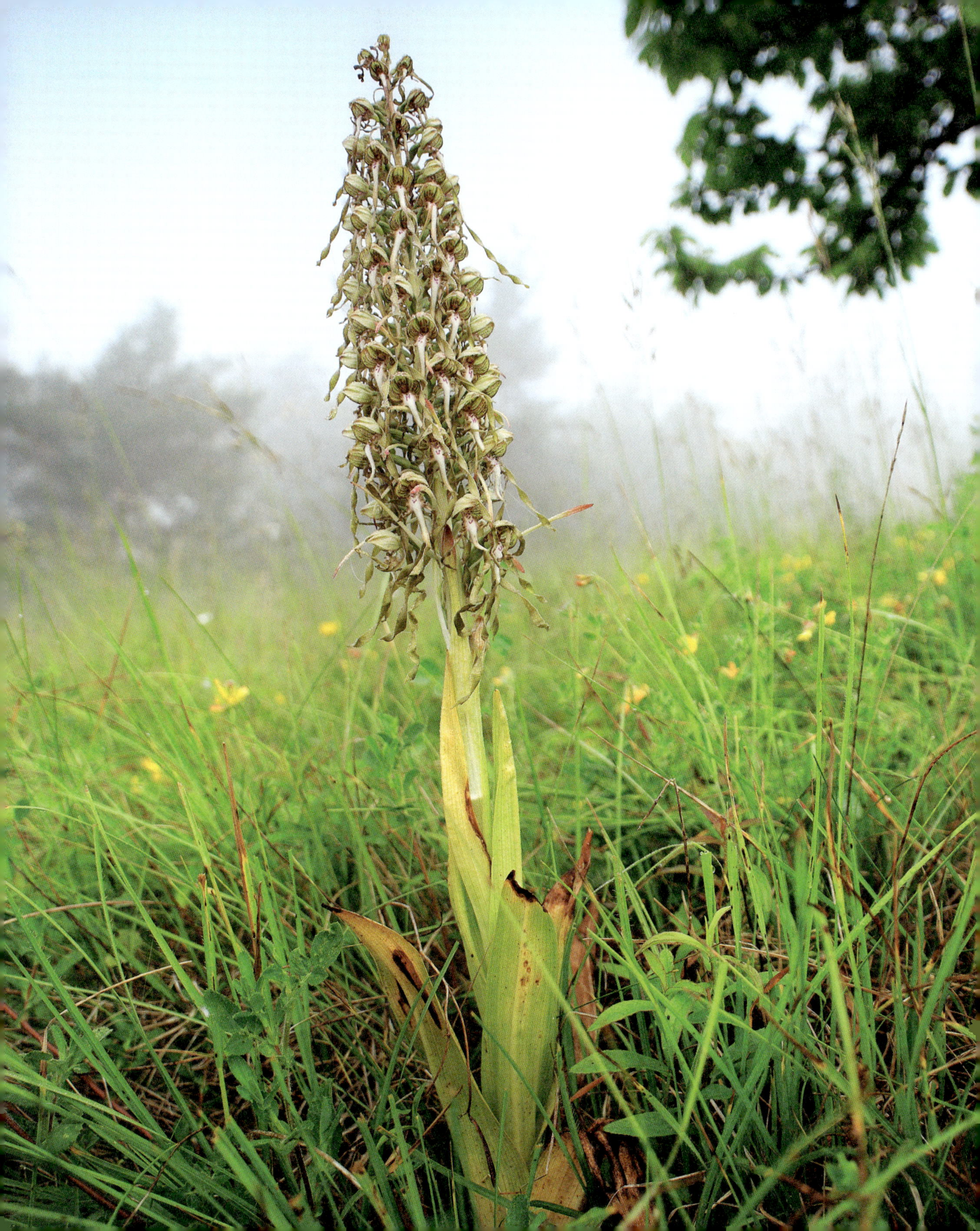

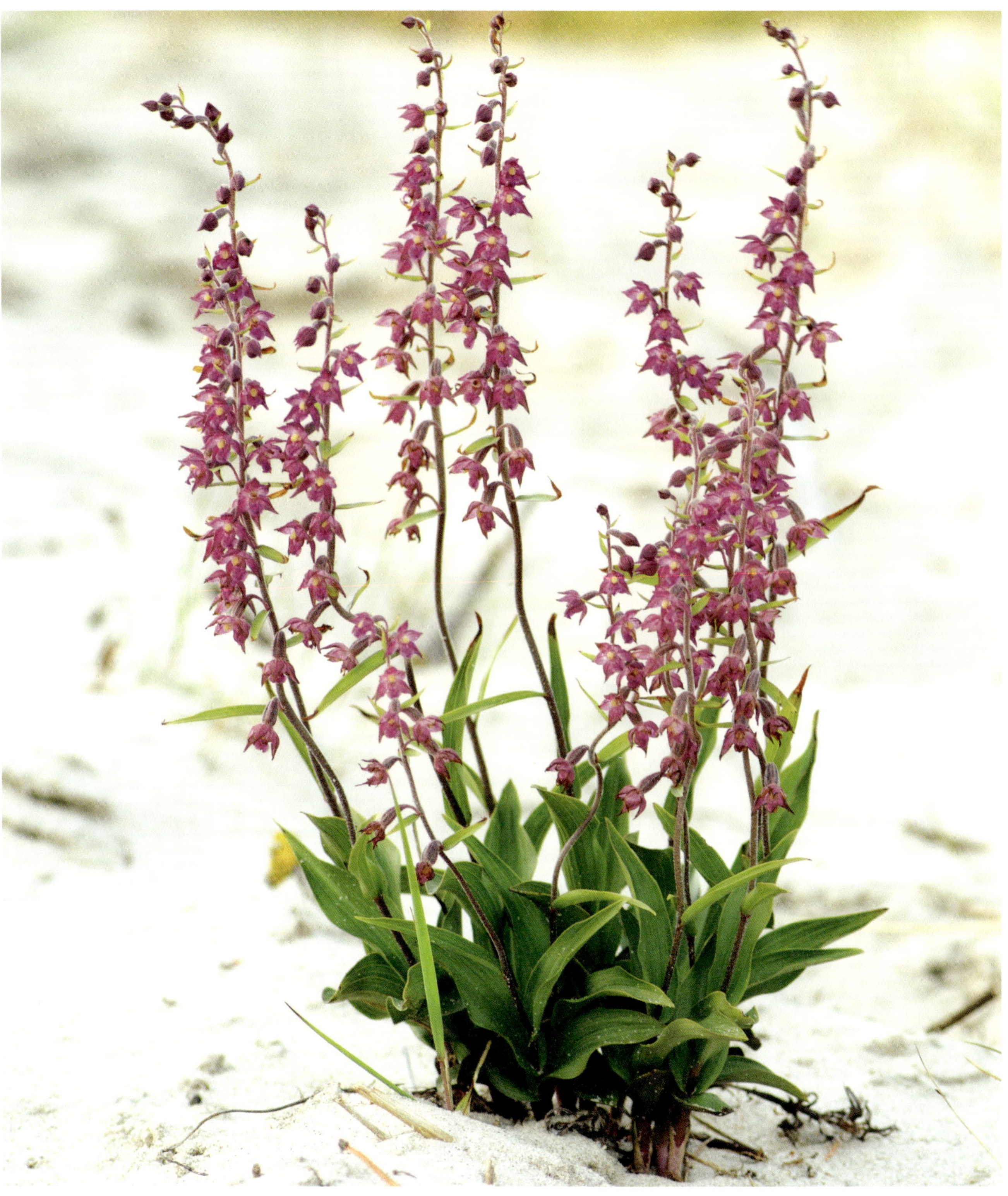

Epipactis helleborine

BROAD-LEAVED HELLEBORINE

THE DESSERT BUFFET IS NOW OPEN

Many species of *Epipactis* are real show-stoppers due to their long stems, some of which can grow to be just short of a yard long. One well-known species is *Epipactis atrorubens*, the dark-red helleborine. It is found in coastal areas, and it has a delightful vanilla scent. The broad-leaved helleborine, on the other hand, owes its success to its sweetness. Instead of merely emitting a deceptive fragrance, it is one of the few orchids that actually rewards its insect pollinator-helpers with a delicious nectar.

STÄNDELWURZ

WILLKOMMEN AM DESSERT-BUFFET

Durch ihren langen, namensgebenden Stängel sind viele Ständelwurz-Arten recht augenfällig, manche werden bis zu 80 Zentimeter hoch. Eine bekannte *Epipactis*-Art ist die an vielen Küsten wachsende Braunrote Ständelwurz (*Epipactis atrorubens*), die sich mit ihrem intensiven Duft auch den Beinamen „Strandvanille" erobert hat. Süß ist ebenso das Erfolgsgeheimnis der Breitblättrigen Ständelwurz: Bestäubenden Insekten bietet sie als eine der wenigen Orchideenarten tatsächlich leckeren Nektar an, anstatt sie nur mit Duft zu täuschen.

Gymnadenia

FRAGRANT ORCHID

SWEET-SMELLING, SPHERICAL BEAUTY

The common name of this orchid says it all, and it truly lives up to its promise, with a delightful fragrance. Overall, *Gymnadenia* species are charmingly sweet, with a strong vanilla scent, such as the "most fragrant" *Gymnadenia odoratissima*. The Chalk fragrant orchid goes a step further by providing an alluring nectar. Nowadays, the *Gymnadenia* genus encompasses species that were previously classified as *Nigritella*, which also produce sweet-smelling cylindrical or spherical flower clusters.

HÄNDELWURZ

KUGELIG SCHÖN UND LIEBLICH DUFTEND

Der Volksname Händelwurz klingt harsch, deutet aber auf die an eine Hand erinnernde Wurzelform hin anstatt auf Zank. Überhaupt besticht die *Gymnadenia* mit Lieblichkeit, nämlich intensivem Vanilleduft wie beispielsweise bei der Wohlriechenden Händelwurz. Die Mücken-Händelwurz legt hier noch eins drauf: Bei ihr gibt es auch Nektar als Lockstoff. Zur Gattung der *Gymnadenia* gehören heute auch die Kohlröschen mit ihrem kugeligen Blütenstand – auch sie duften süß und zum Glück nicht nach Kohl!

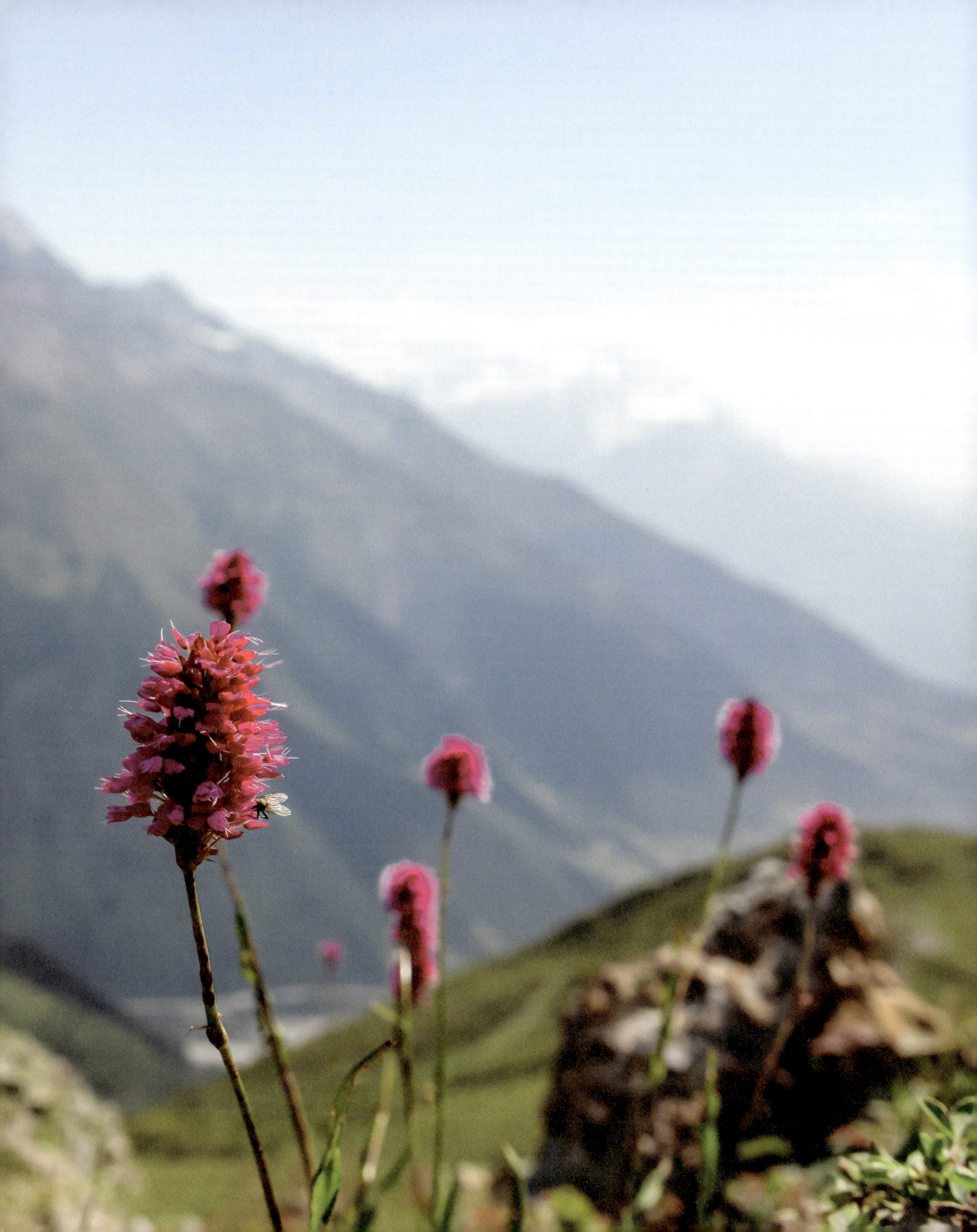

Endangered species

LADY'S SLIPPER AND BROAD-LEAVED MARSH ORCHIDS

CYPRIPEDIUM CALCEOLUS AND *DACTYLORHIZA MAJALIS* MAY BE BEAUTIFUL, BUT THEY ARE IN DISTRESS

Only a century ago, the lady's slipper thrived in forest clearings. But poor environmental conditions have taken their toll, with *Cypripedium calceolus* now in peril. In a given area where a five-score of these orchids once grew, statistically speaking there are only three plants left. The causes include intensive forest management, over-fertilization, and the effects of climate change, which have dried out once-moist habitats. Wild orchids are also at risk from direct human disturbances. The broad-leaved marsh orchid, *Dactylorhiza majalis*, is an alarming example. It grows mainly in moist meadows, but it has been pushed out of its former habitat, too much of which was valued as land to be drained and converted to farming and development. Even where these increasingly rare orchids still grow, they are in danger from reckless orchid "fans" who dig up or trample the plants in pursuit of the perfect photo. For these reasons, the lady's slipper and broad-leaved marsh orchid are now endangered. Conservationists are trying to purchase threatened habitat and engaging in advocacy to save what remains from wanton destruction.

Gefährdete Arten

FRAUENSCHUH UND KNABENKRAUT

CYPRIPEDIUM CALCEOLUS UND *DACTYLORHIZA MAJALIS*: WUNDERSCHÖN – UND GANZ SCHÖN IN BEDRÄNGNIS

Noch vor 100 Jahren war der Gelbe Frauenschuh ein prominenter Bewohner von Waldlichtungen – doch die immer schlechteren Umweltbedingungen haben *Cypripedium calceolus* in Bedrängnis gebracht. Wo einst 100 Frauenschuhe wuchsen, finden sich heutzutage nur noch drei Pflanzen. Ursachen sind intensive Forstbewirtschaftung, Überdüngung und die Folgen des Klimawandels, die einst feuchte Wachstumsgebiete austrocknen lassen. Doch der Mensch stört noch viel aktiver die Lebensräume der Wildorchideen – ein alarmierendes Beispiel ist *Dactylorhiza majalis*: Das Breitblättrige Knabenkraut wächst vornehmlich in feuchten Wiesen. Da aber Acker- und Bauland wertvoller sind, wurde dieser Lebensraum in den vergangenen Jahrzehnten durch Entwässerung und Bebauung skrupellos zurückgedrängt. Und dort, wo die immer seltener werdenden Orchideen noch wachsen, sind sie ausgerechnet durch rücksichtslose „Orchideenfans" in Gefahr, die die Pflanzen ausgraben oder auf der Jagd nach Fotos niedertrampeln.

Heute stehen Frauenschuh und Breitblättriges Knabenkraut auf der Roten Liste der gefährdeten Arten. Daher versuchen Naturschützer inzwischen, gezielt gefährdete Gebiete aufzukaufen und mit Aufklärungskampagnen zu verhindern, dass noch mehr Lebensraum zerstört wird.

NEXT PAGE | *Cypripedium calceolus* (left) and *Dactylorhiza majalis* (right)
NÄCHSTE SEITE | Frauenschuh (links) und Knabenkraut (rechts)

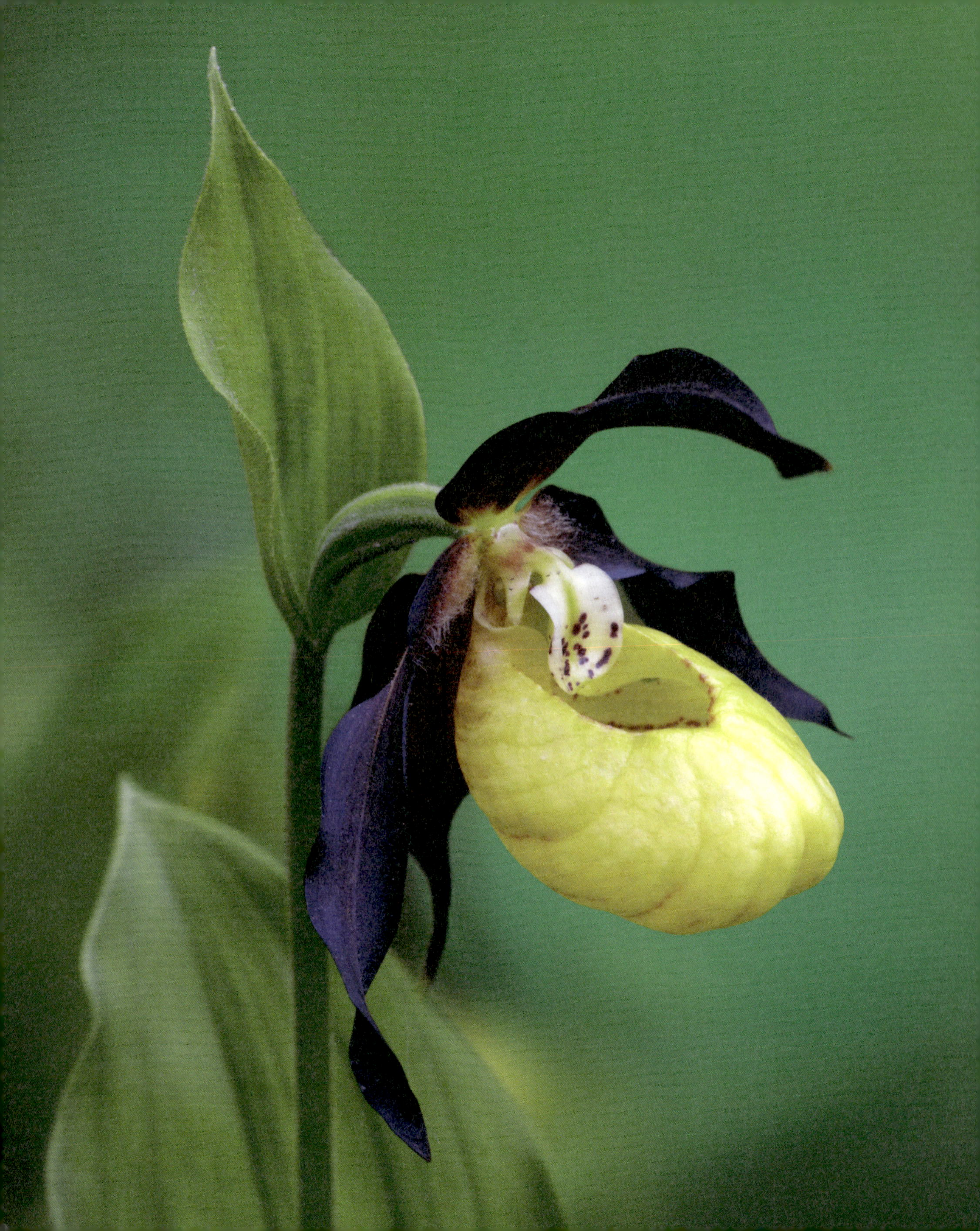

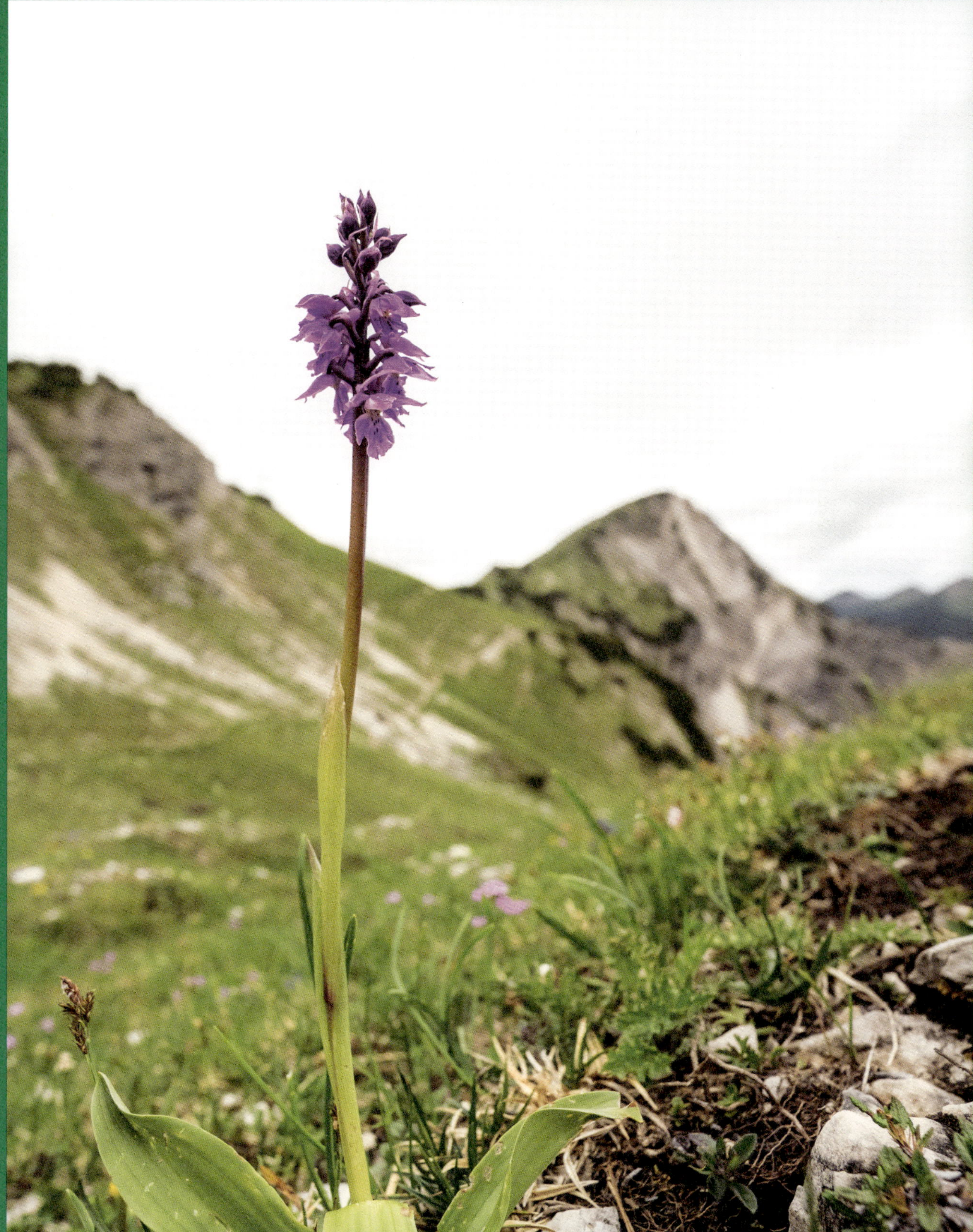

Thuringia's oldest nature reserve

THE ORCHIDS IN THE VALLEY

THE LEUTRA VALLEY NEAR JENA, GERMANY, IS A FLOURISHING PARADISE DESPITE CONSTANT STRUGGLE

Central European terrestrial orchids thrive in barren soils like the shell limestone found in the region around Jena. The Leutra Valley is home to up to twenty-seven species of orchids, including the frog orchid and the greater butterfly orchid, which flourish in alkaline meadows and on sparsely vegetated slopes. The local nature conservation association offers guided hikes for orchid lovers during their flowering season to prevent ignorant orchid "fans" from wrecking these protected natural treasures.

Unfortunately, the habitat of wild orchids in the Leutra Valley has been threatened for a long time. In the 1930s, the valley was declared a nature reserve – even as they pushed an autobahn through it at the same time. Reportedly, many orchids survived only because road construction workers took it upon themselves to dig them up and transplant them into nearby meadows. Since the dismantling of the autobahn in the mid-2010s, rare orchids have been reclaiming their native habitats in this valley. Many, such as the lizard orchid, can now be hiked out to and observed in woods and meadows – with consideration and knowledgeable guides rather than noise and exhaust fumes.

NEXT PAGE | The Leutra Valley near Jena (left),
one of its inhabitants: a marsh orchid (right)

Thüringens ältestes Naturschutzgebiet

IM TAL DER ORCHIDEEN

IM LEUTRATAL BEI JENA FINDET SICH EIN BLÜHENDES PARADIES – UND KÄMPFT SEIT JEHER MIT WIDRIGKEITEN

Mitteleuropäische Erdorchideen lieben karge Böden wie den Muschelkalk, den es in der Region um Jena gibt. Der Artenreichtum im Leutratal südlich der Großstadt sucht in Deutschland seinesgleichen: Bis zu 27 Orchideenarten vom Blassen Knabenkraut bis zum Großen Zweiblatt wachsen auf den kalkigen Wiesen und lichten Hängen. Der örtliche Naturschutzbund bietet zur Blütezeit geführte Wanderungen für Orchideenfreunde an – wichtig, um zu verhindern, dass unwissende Orchideenfans im Entdeckerübermut die geschützten Naturschätze zerstören.

Dass dem Lebensraum der wilden Orchideen Gefahren drohen, weiß man im Leutratal leider schon lange: In den 1930er-Jahren wurde, gleichzeitig mit der Erklärung zum Naturschutzgebiet, eine Autobahn durchs Tal gebaut! Man berichtet, dass das Überleben vieler Exemplare nur rührigen Straßenbauarbeitern zu verdanken ist: Sie gruben die Orchideen aus und verpflanzen sie in etwas entfernte Wiesenstücke. Seit dem Rückbau der Trasse Mitte der 2010er-Jahre erobern sich die heimischen Orchideenraritäten ihren Lebensraum zurück: Eine Vielzahl an Knabenkräutern, Bocks-Riemenzunge & Co. können nun wieder ohne Lärm und Abgase in den Wiesen und Waldstücken erwandert werden – natürlich mit Rücksicht und fachkundiger Führung.

NÄCHSTE SEITE | Das Leutratal bei Jena (links) und einer seiner Bewohner: ein Knabenkraut (rechts)

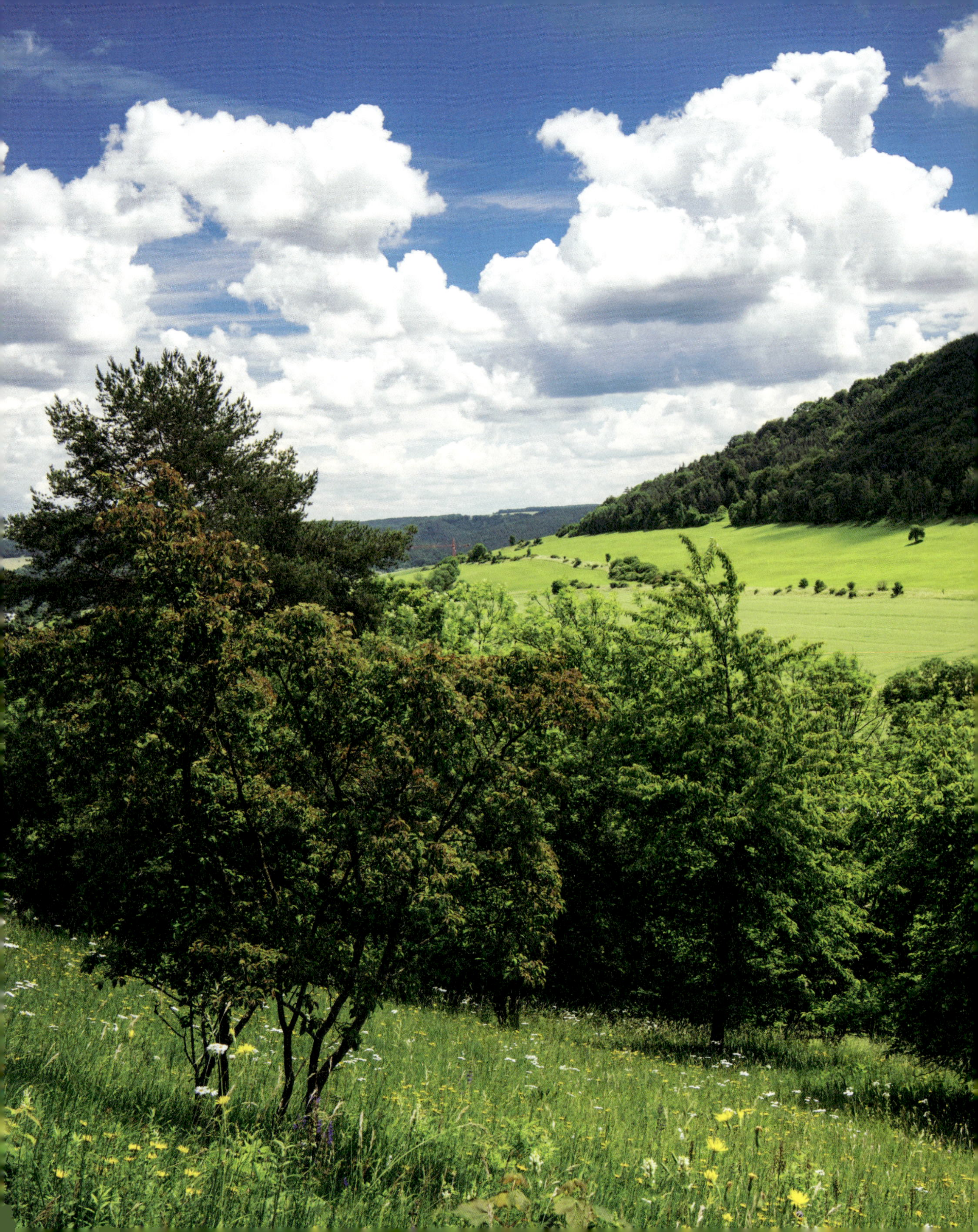

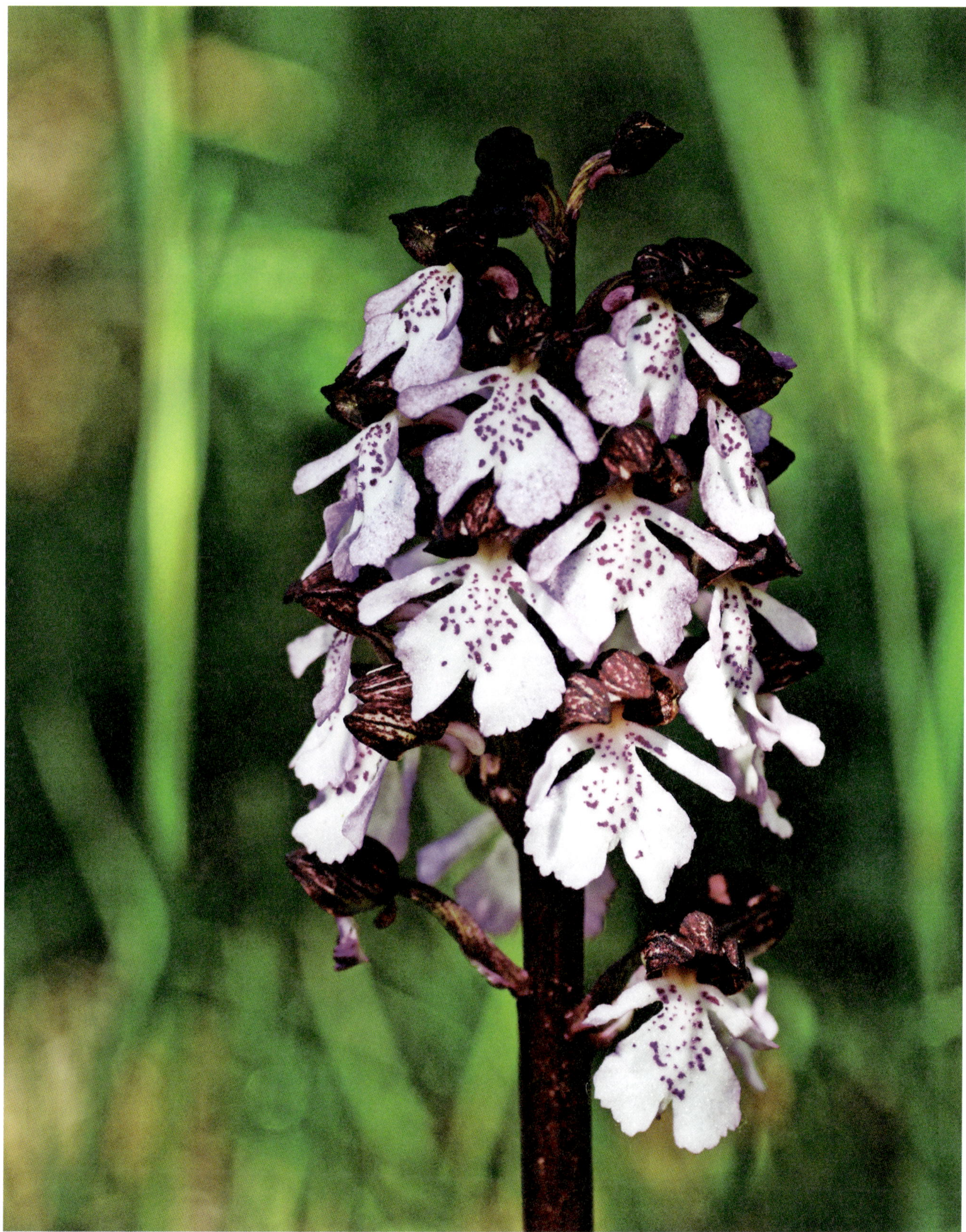

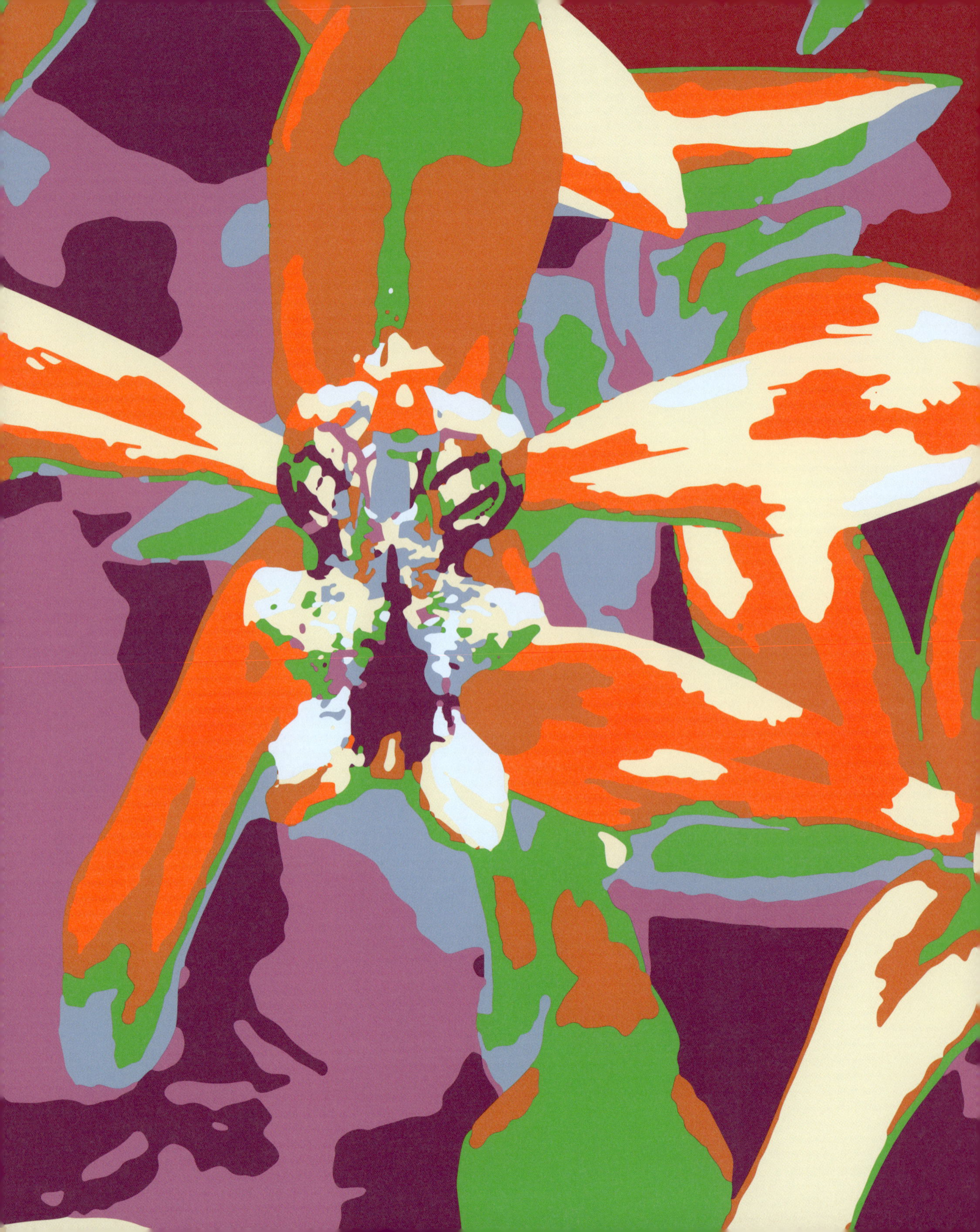

FASCINATING BOTANY

FASZINIERENDE BOTANIK

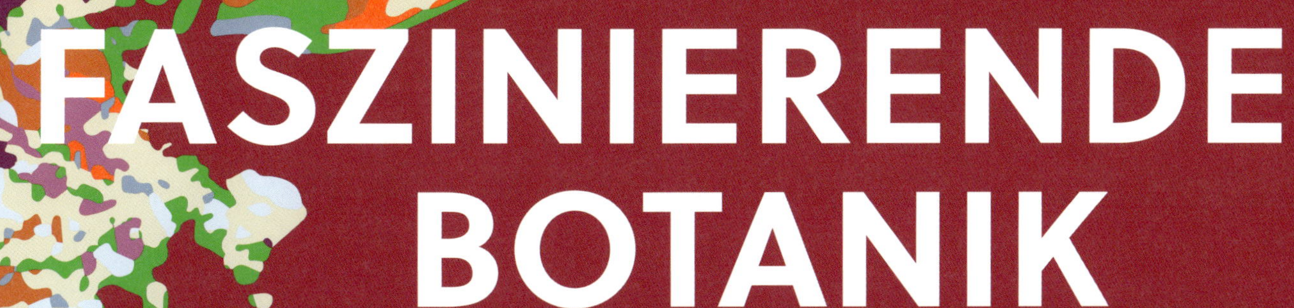

Fungal survival aids

CRUCIAL ALLIES

FROM GERM TO FLOWER, ORCHIDS OWE THEIR EXISTENCE TO MYCORRHIZA

Mycorrhiza refers to a special relationship between plants and fungi. For mature orchids, the fungi help provide them with the minerals and water they need to grow and thrive. In return, the plant shares with the fungi some of the nutrients it produces through photosynthesis. Interestingly, the fungi that settle on the roots of orchids are usually specific to the species of plant they're associated with. That's why if you try to grow an orchid that's been taken from the wild, it might not thrive in your garden – it needs the right kind of fungus to survive. But it's not just about mature orchids. The fungi also play a critical role in helping orchid seeds germinate and grow. The seeds of orchids are tiny and lack the nutrients they require, but the fungi provide them with a "starter package", letting them grow into healthy seedlings.

Mycorrhizal fungi aren't only found in forest soil, either. They are also important partners for epiphytic tropical orchids. In these cases, the fungi settle on the silvery velamen layer of the orchid's aerial roots and sometimes even penetrate the bark of the host plant. This allows orchid seeds to germinate and young plants to grow successfully, at lofty heights and in the lee of an adult plant.

Pilze als Überlebenshelfer

WICHTIGSTE VERBÜNDETE

ERST MYKORRHIZA ERMÖGLICHEN DEN ORCHIDEEN IHRE EXISTENZ – VOM KEIM BIS ZUR BLÜTE

Mit Mykorrhiza wird eine Symbiose von Pflanzen und Pilzen bezeichnet. Bei Orchideen ist es so: Den erwachsenen Orchideen liefern Pilze Mineralstoffe und Wasser. Die Pflanze wiederum versorgt den Pilz aus ihrer Photosynthese mit Nährstoffen. Die auf den Wurzeln angesiedelten Mykorrhizapilze sind übrigens meist spezifisch auf die jeweilige Pflanzenart ausgerichtet – daher lassen sich aus der Natur geraubte Orchideen nicht einfach so im eigenen Garten kultivieren! Auch ist es der Pilz, der das Auskeimen der Orchideensamen erst möglich macht, und der Keimling kann nur durch die Mykorrhizaversorgung überleben, da die winzigen Samenkörnchen selbst kein „Nährstoff-Starterpaket" an Bord haben.

Mykorrhizapilze sind übrigens nicht nur im Waldboden angesiedelt. Auch für epiphytisch wachsende Tropenorchideen sind sie wichtige Partner: Dort sitzen sie auf der silbrigen Velamen-Schicht der Luftwurzeln und dringen sogar in die Baumrinde ihrer Gastgeberpflanzen ein. So wird es möglich, dass auch in luftigen Höhen, quasi im Windschatten von erwachsenen Pflanzen, Orchideensamen keimen und Jungpflanzen erfolgreich anwachsen können.

Mimicry

THE ORCHIDS AND THEIR BAG OF TRICKS

A FRAGRANCE USEFUL AND SEDUCTIVE

Orchids have truly become masters of deception. They give off a sweet scent to lure food-seeking insects into their flowers. But in most cases, the insects find neither nourishing pollen nor nectar. And so, to satisfy their hunger, they move on – to the next flower. The next flower is then pollinated with the pollen that has stuck to the insect. Interestingly, some orchid species emit their alluring scent only at certain times of the day, precisely when the pollinators they are targeting are on the move.

Some orchid species rely on an insect's reproductive instincts rather than on its desire to feed. The orchids attract these pollinators by imitating the sex pheromones of certain insect species, often combined with a visual attraction. For example, the bee orchid, the late spider orchid, and the fly orchid use their scents to lure specific male pollinators. In the case of the bucket orchid, *Coryanthes*, the plant plays "wing man" for male bees by providing them with a kind of "manly" scent that makes them more attractive to the females.

RIGHT PAGE | The bucket orchid provides male bees with a scent that makes them more attractive to the females.

RECHTE SEITE | Die Orchidee der Gattung *Coryanthes* versorgt Bienenmännchen mit einem Duft, der diese für Weibchen attraktiver macht.

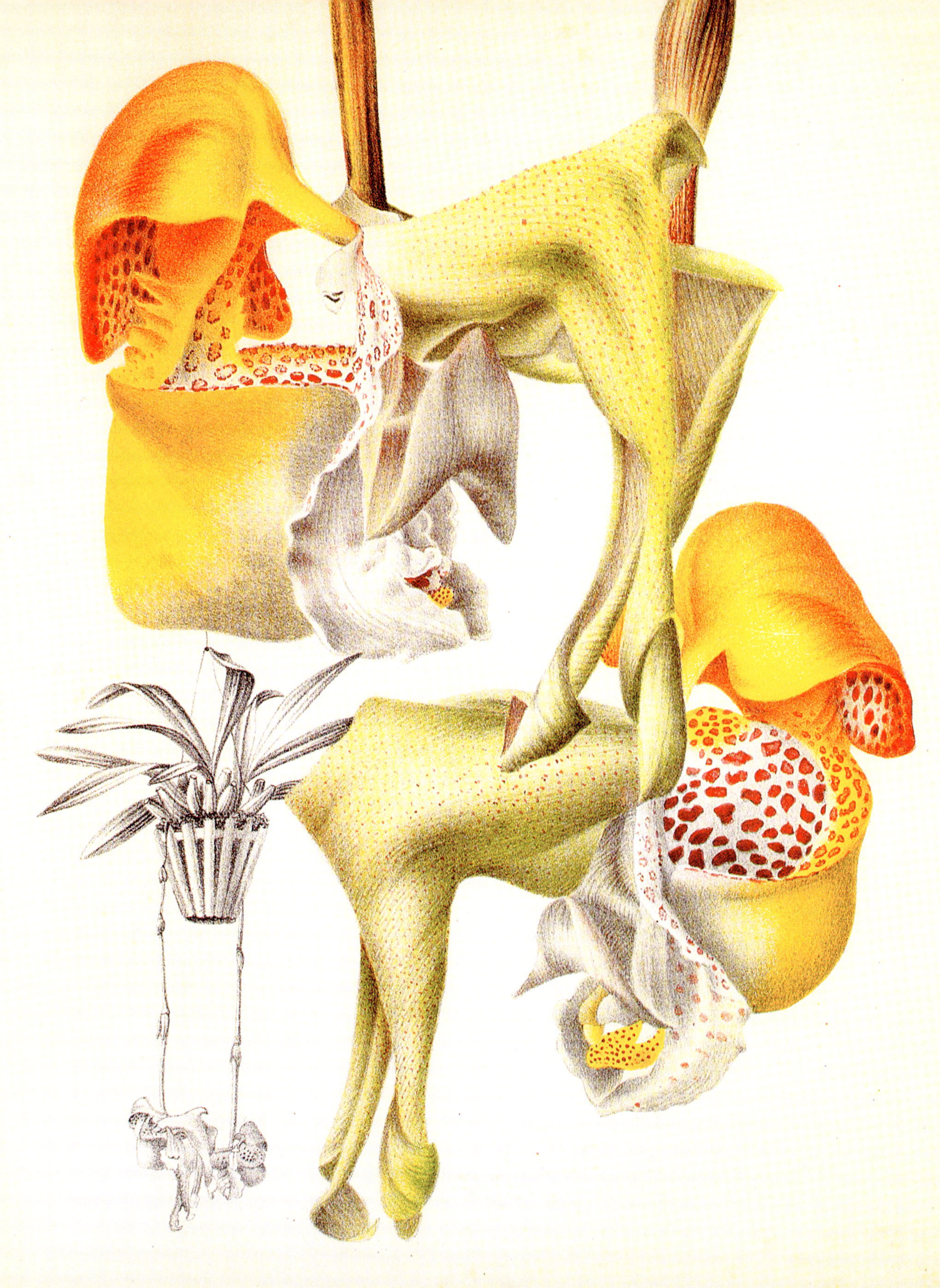

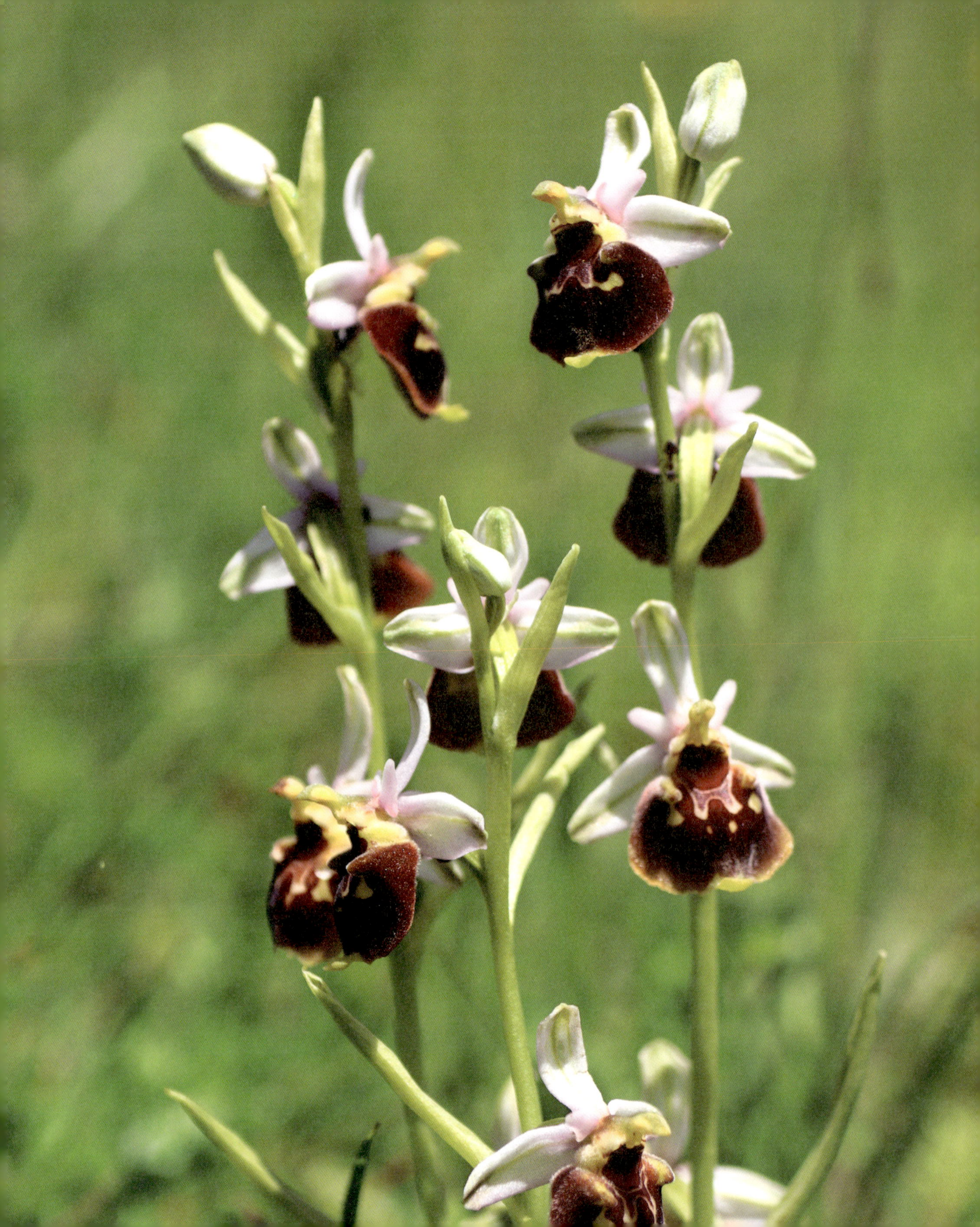

Mimikry

DIE TRICKKISTE DER ORCHIDEEN

BETÖRENDER DUFT, ATTRAKTIVES AUSSEHEN

Orchideen haben sich zu wahren Meistern der Täuschung entwickelt. So verströmen sie zum Beispiel einen süßen Duft, der nahrungssuchende Insekten in die Blüte krabbeln lässt. Dort finden sie in den allermeisten Fällen aber weder nährende Pollen oder Nektar vor – also nichts wie auf zur nächsten Blüte, um den Hunger zu stillen! Mit den Pollen, die dabei an den Insekten hängenbleiben, wird dann die nächste Blüte bestäubt. Interessant: Ihren verlockenden Duft verströmen einige Orchideenarten nur zu bestimmten Tageszeiten – nämlich genau dann, wenn die für sie interessanten Bestäuber unterwegs sind.

Manche Orchideenarten setzen nicht auf den Fress-, sondern den Fortpflanzungstrieb: Sie imitieren mit ihrem Duft Sexuallockstoffe bestimmter Insektenarten, oft gepaart mit einem optischen Reiz. Bei einem Blick auf die Blüte von Bienen-Ragwurz, Hummel-Ragwurz oder Fliegen-Ragwurz wird schnell klar, welche Männchen der jeweiligen Orchidee in die Bestäubungsfalle gehen sollen. Dabei ist die Pflanze aber auch der „Flügelmann" des Insektenmännchens, wie z. B. im Fall der Eimer-Orchidee *Coryanthes*: Sie versorgt die Bienenmännchen mit einer Art „Herrenduft", der diese wiederum für Weibchen attraktiver macht.

LEFT PAGE | Despite its name, the Late spider orchid is mainly pollinated by bees.

LINKE SEITE | Der Name täuscht ein wenig: Die Weiße Hummel-Ragwurz wird tatsächlich in der Hauptsache von Bienen bestäubt.

Cypripedium and Pterostylis

IT'S A TRAP!

ORCHIDS LURE INSECTS WITH SLIPPERY SLIDES AND ANIMATED FLOWERS

It seems an orchid will do anything to reproduce. The lady's slipper orchid, whose "slipper" petal is an ingenious, slippery booby trap, does it very cleverly. With its striking colors and scents, it attracts a pollinator insect. As soon as it settles onto the flower, the credulous insect is doomed to slide farther in on an oil-like film. There is no going back on the smooth walls of the blossom, and the insect's only escape is through the back door. But first it must squeeze past the stigma, where any pollen the insect already has on it rubs off before it picks up more along the way. This complicated, kettle-trap mechanism lets *Cypripedium* avoid self-pollination. One could even say it's a protective instinct of nature to develop healthy genetic material.

Compared to the lady's slipper orchid, though, the Australian genus *Pterostylis* – the greenhood orchids – seem downright aggressive. In these species, the central petal, or lip, is movable. And so these orchids look like miniature predators. If an insect is attracted to the flower and touches the lip, the lip springs forward, the fly is flung into the flower, and then the lip closes upon it, accomplishing a forced pollination.

NEXT PAGE | Left *Cypripedium*, right a *Pterostylis* from Australia. Both ensnare insects that can only escape by squeezing past the stigma, where they pollinate the orchid. The pollinators usually emerge unscathed.

Cypripedium und Pterostylis

VORSICHT, FALLE!

MIT GLITSCHIGEN GLEITFALLEN UND ZIEMLICH LEBENDIGEN BLÜTEN ZWINGEN ORCHIDEEN INSEKTEN DAZU, SIE ZU BESTÄUBEN

Was Orchideen nicht alles anstellen, um ihren Fortbestand zu sichern! Ganz raffiniert macht es der Frauenschuh, dessen „Pantoffel"-Blütenblatt eine ausgeklügelte Gleit- oder Kesselfalle ist. Mit auffälligen Farben und Duftstoffen wird ein Insekt angelockt, und hat es sich erstmal unvorsichtigerweise auf der betörenden Blüte niedergelassen, ist es schon zu spät: Auf einem ölartigen Film rutscht es tief in die Falle hinein. Da es an den glatten Blütenwänden den Rückweg nicht mehr antreten kann, bleibt nur noch die Flucht durch die „Hintertür". Auf diesem Weg quetscht sich das Insekt erst an der Narbe vorbei, wo bereits mitgebrachte Pollen kleben bleiben, ehe es auf dem weiteren Weg dann neue Pollen aufnimmt. Mit diesem komplizierten Mechanismus verhindert die Kesselfalle bei *Cypripedium* die Selbstbestäubung – gewissermaßen ein Schutzinstinkt der Natur, um gesundes Erbgut zu entwickeln.

Fast schon aggressiv wirkt im Vergleich zur Kesselfalle die australische *Pterostylis*, auch Grünkappe genannt: Durch deren bewegliches zentrales Blütenblatt, die Lippe, wirkt die Orchidee wie ein kleines Raubtier. Berührt ein angelocktes Insekt diese Lippe, schnellt sie vor, schleudert die Fliege in die Blüte und verschließt sie anschließend, um eine Bestäubung zu erzwingen.

NÄCHSTE SEITE | Links *Cypripedium*, rechts die australische *Pterostylis*. Beide ‚fangen' Insekten ein, die sich bei der Flucht draußen an der Narbe vorbeizwängen müssen und die Orchidee so bestäuben. Zu Tode kommen die Insekten übrigens in der Regel nicht.

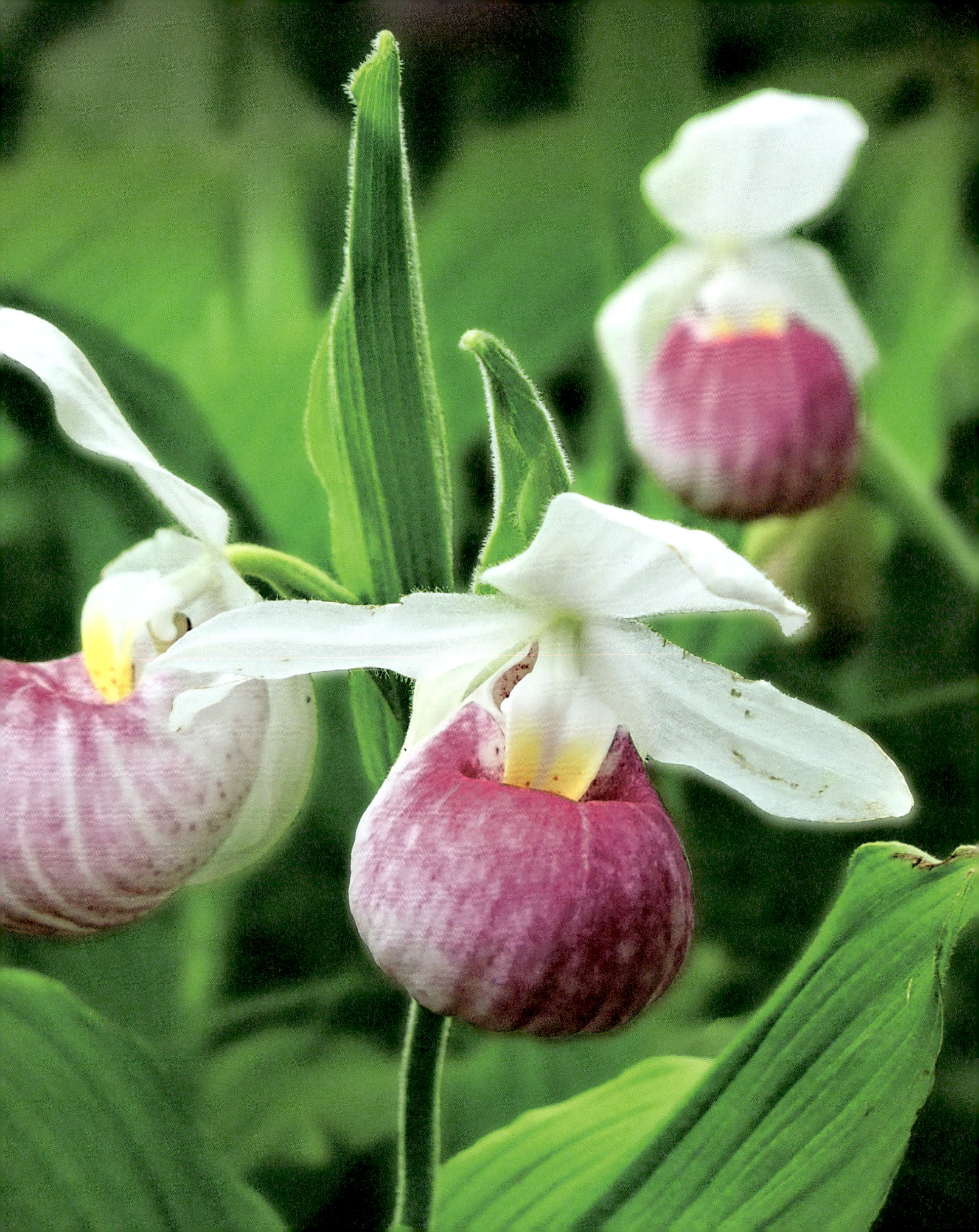

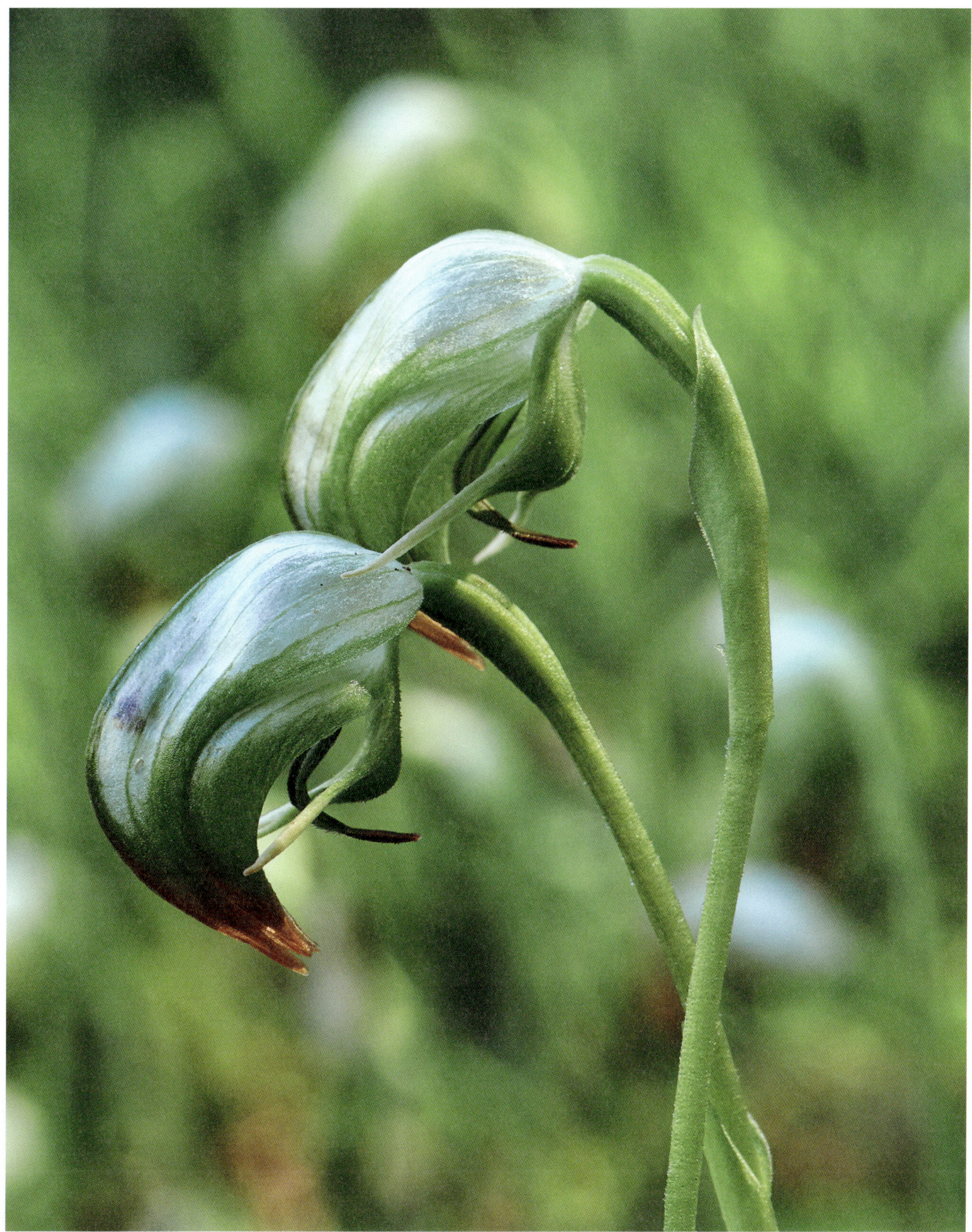

Angraecum sesquipedale

THE ORCHID AND THE MOTH

WITH A KEEN INTUITION FOR INSECTS, DARWIN PREDICTS A BUTTERFLY DECADES BEFORE ITS ACTUAL DISCOVERY

The peculiar shapes of certain orchid flowers can sometimes make you wonder what nature is up to. Charles Darwin experienced this in 1862 when he observed the white flowers of the Star of Madagascar *(Angraecum sesquipedale)*. He discovered that the flower's nectar spur, which was almost a foot long, contained nectar deep down at the bottom. Darwin, quite the sleuth, realized that the concealed nectar and unusual shape of the flower only made sense if there was an insect with an equally long proboscis. The only thing was, no insect capable of pollinating the orchid had ever been discovered.

And lo and behold, some forty years later, in 1903, a moth was discovered on Madagascar that matched the description of this phantom that Darwin had referred to: a sphinx moth that, with a foot-long proboscis, could access the nectar of the Star of Madagascar. To pay tribute to Darwin's fascinating prediction, this subspecies of the long-tongued moth was named *Xanthopan morganii praedicta*, "the predicted one." It is worth noting that almost a century passed before it was possible to properly document how the moth is able to sip nectar from the orchid's flower. Darwin's remarkable intuition for insects never ceases to astonish.

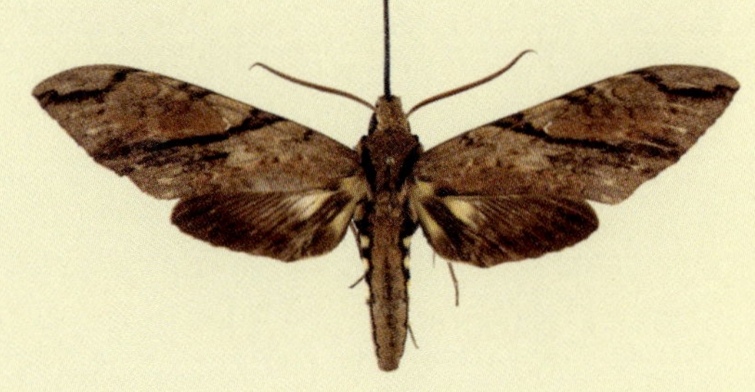

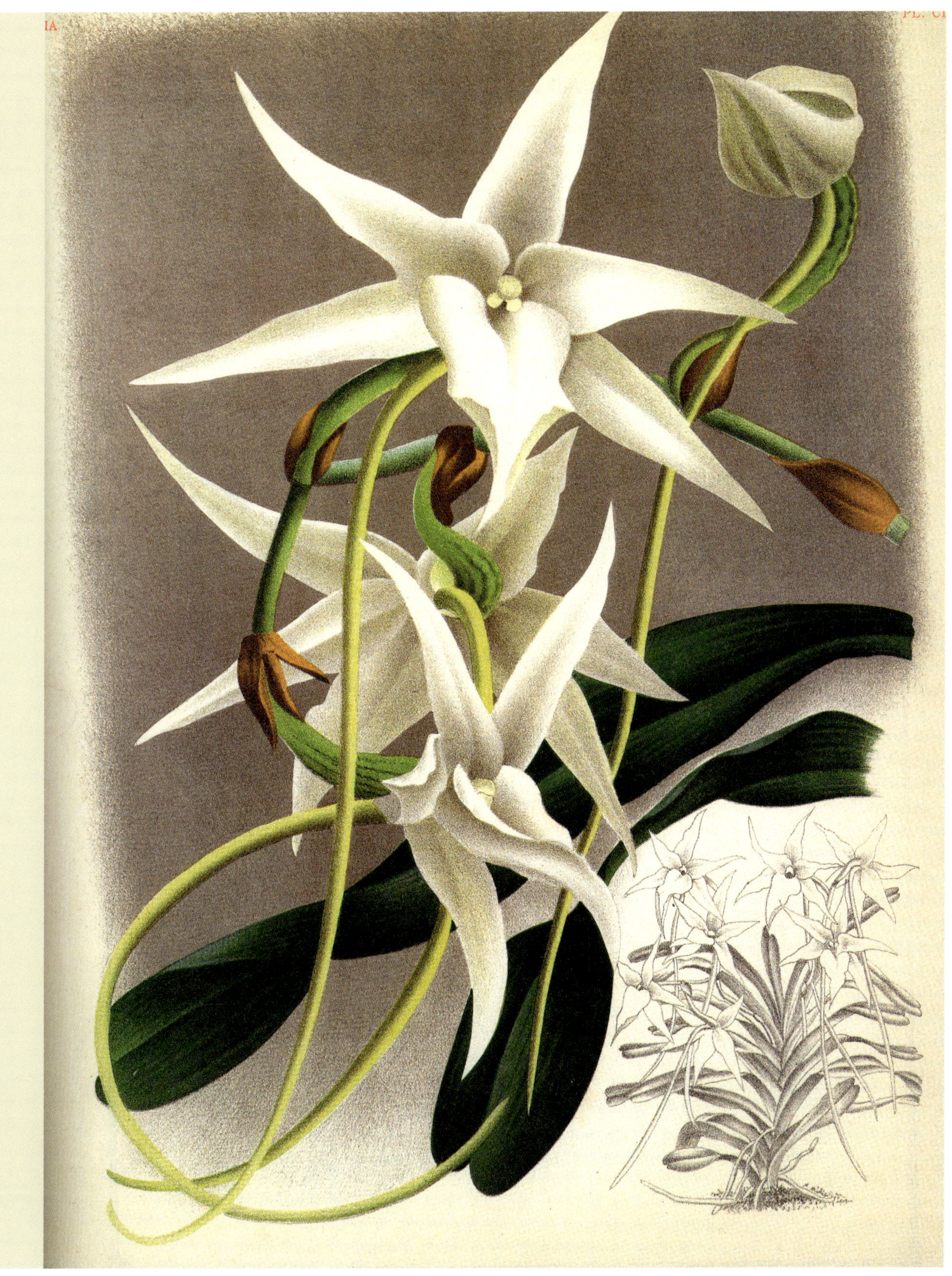

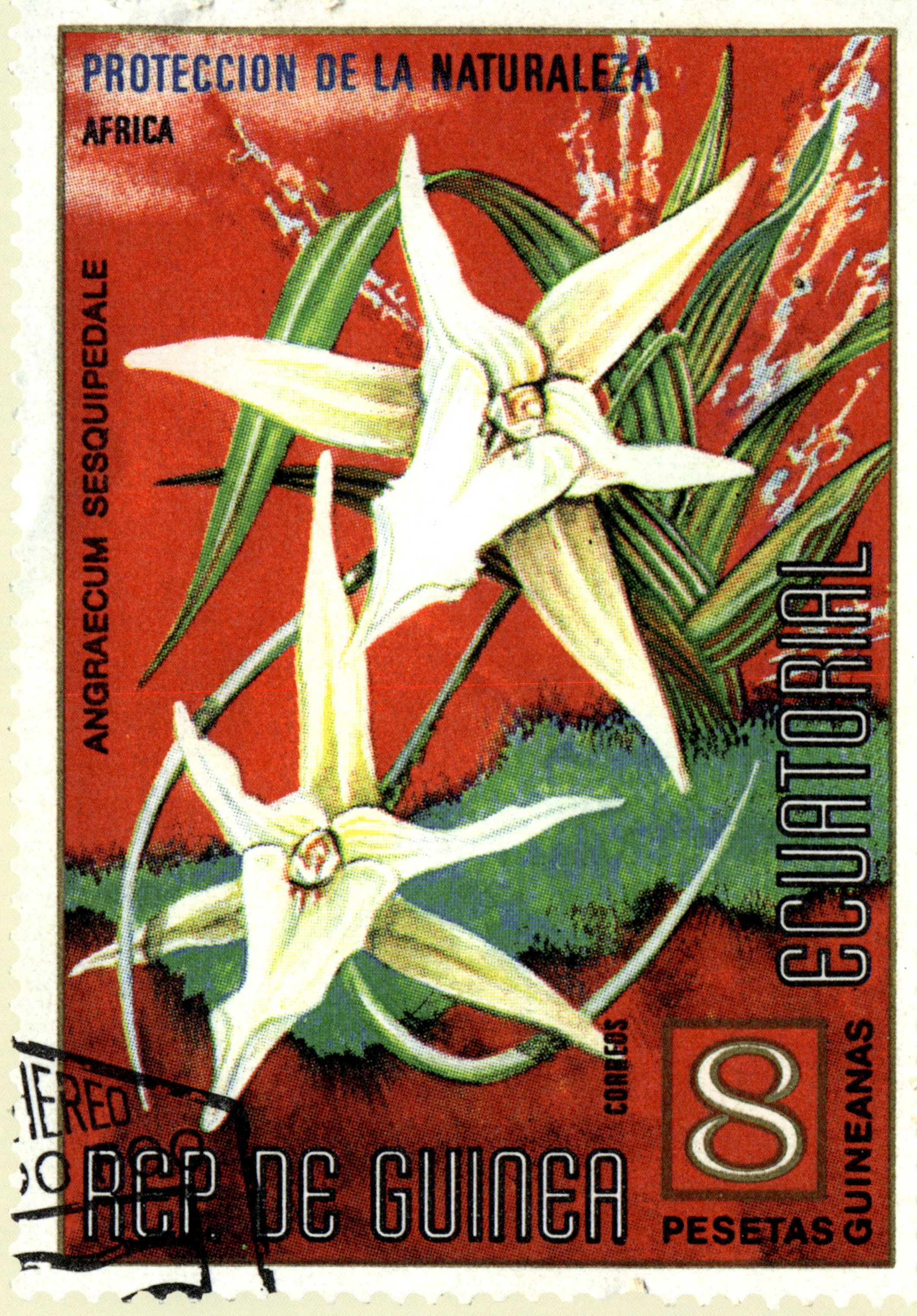

PROTECCION DE LA NATURALEZA
AFRICA
ANGRAECUM SESQUIPEDALE
ECUATORIAL
CORREOS
8
PESETAS
GUINEANAS
REP. DE GUINEA

Angraecum sesquipedale

DIE ORCHIDEE UND DER NACHTFALTER

DARWINS UNTRÜGLICHES GESPÜR FÜR INSEKTEN LÄSST IHN EINEN SCHMETTERLING VORHERSAGEN, DER ERST JAHRZEHNTE SPÄTER ENTDECKT WIRD

Die bizarren Formen mancher Orchideenblüten werfen bisweilen die Frage auf, was sich die Natur wohl dabei gedacht haben mag. So ging es auch Charles Darwin: Als er 1862 die weißen Blüten des Sterns von Madagaskar (*Angraecum sesquipedale*) betrachtete, stellte er fest, dass der rund einen Fuß lange Blütensporn ganz am unteren Ende mit Nektar gefüllt war. Darwins These zeigt detektivischen Spürsinn: Damit der versteckte Nektar und die bizarre Blütenform Sinn ergeben, muss es ein Insekt mit ebenso langem Rüssel geben. Das Problem: Niemand hatte bislang ein solches Insekt gesehen, welches anatomisch in der Lage gewesen wäre, die Orchidee zu bestäuben.

Und tatsächlich: 1903, also erst rund 40 Jahre später, wurde auf Madagaskar ein Nachtfalter entdeckt, der genau auf Darwins „Phantombeschreibung" passte: Wenn die Sphinx-Motte ihren einen Fuß langen Saugrüssel entrollt, ist der Nektar des Sterns von Madagaskar in Reichweite. Zu Ehren der faszinierenden Voraussage wurde die Nachtfalter-Unterart Xanthopan morganii praedicta – „die Vorausgesagte" getauft. Übrigens konnte erst ein knappes Jahrhundert später wirklich dokumentiert werden, wie sich der Nachtfalter den Nektar aus der Orchideenblüte angelt – und Darwins untrügliches Gespür für Insekten ist noch immer verblüffend.

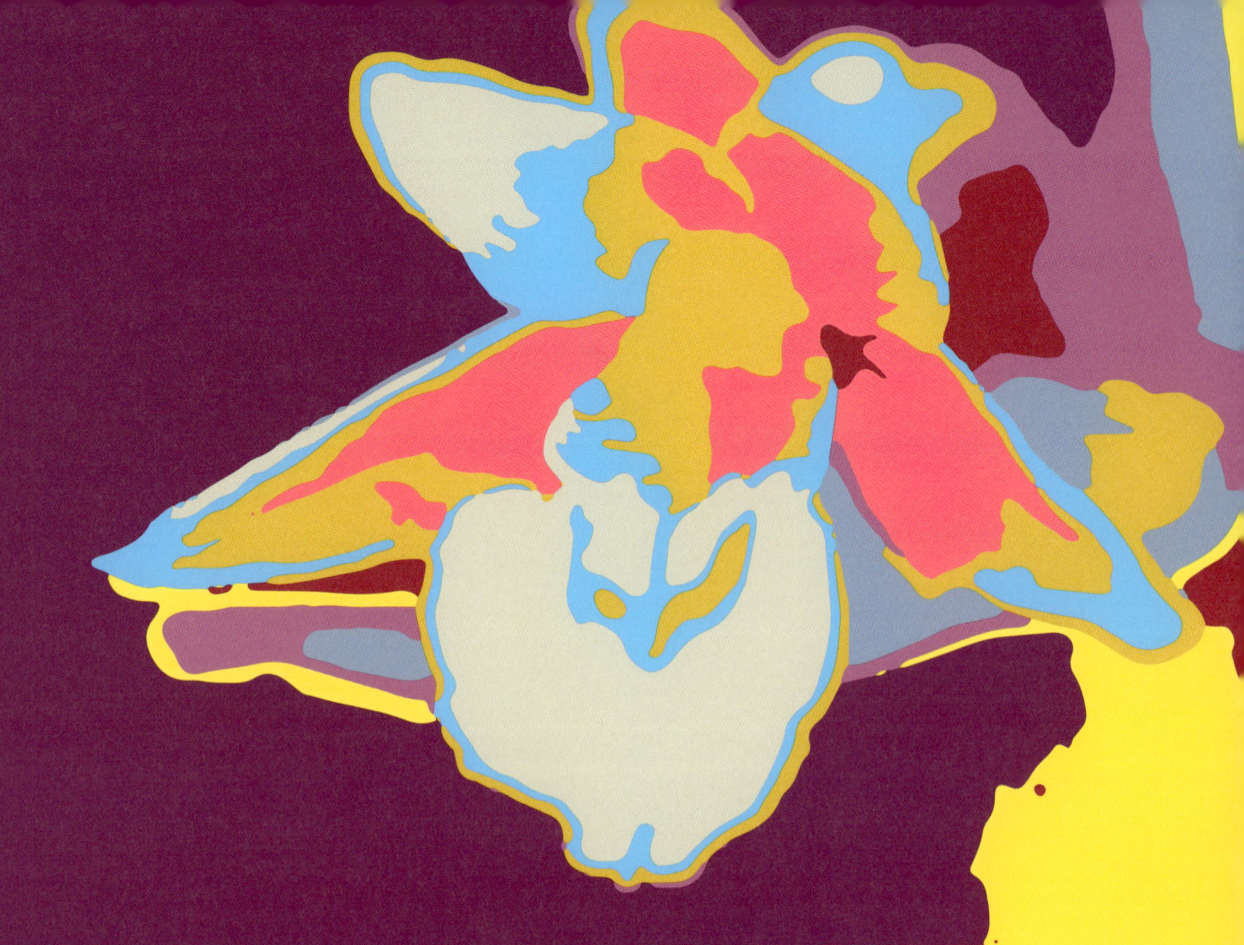

MASTERPIECES OF NATURE

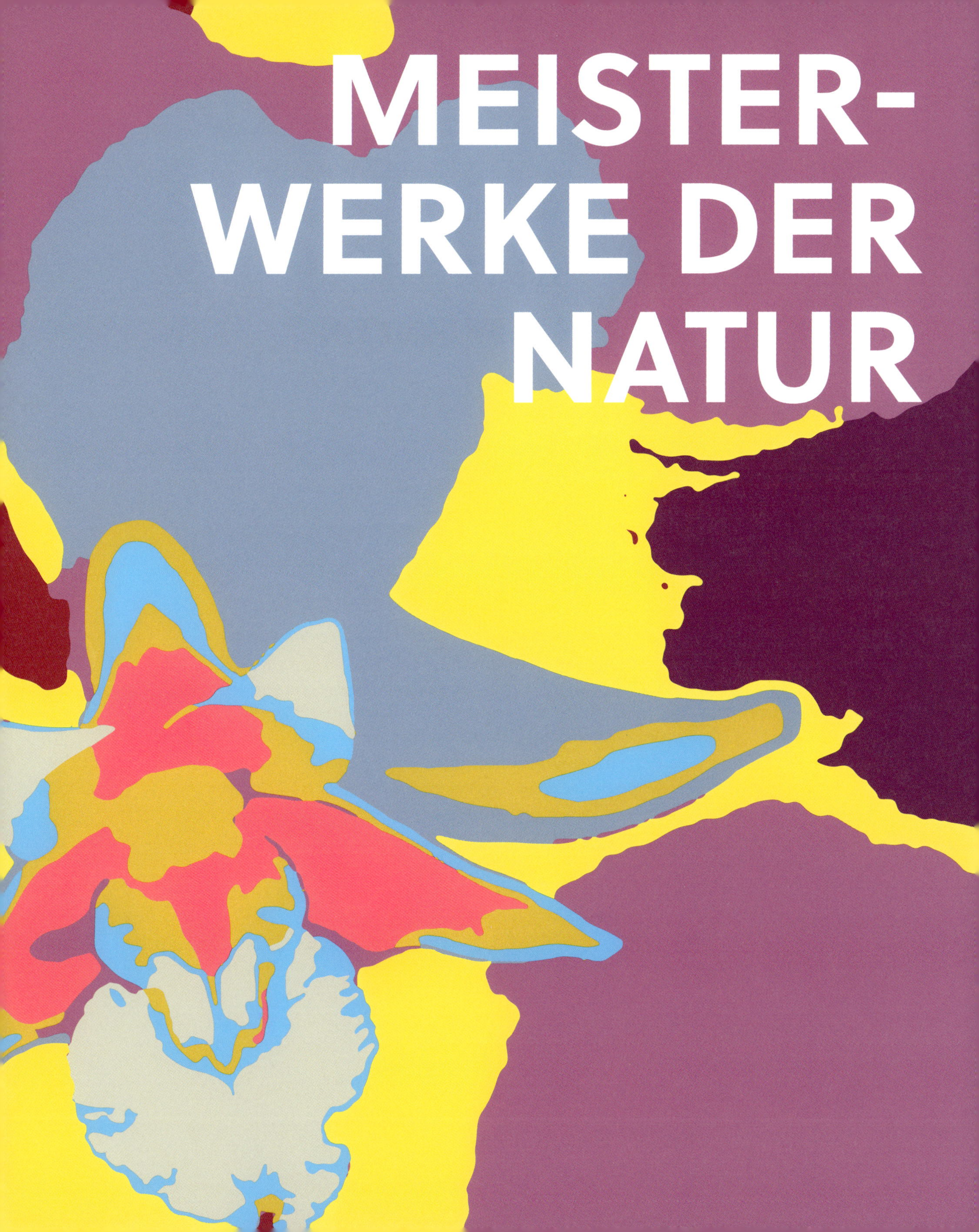
MEISTER-
WERKE DER
NATUR

Orchids with faces

THE *PAREIDOLIA* PHENOMENON

A PSYCHOLOGICAL ACCOUNT OF WHY WE SEE FACES IN THINGS

As humans, we tend to be creatures of habit, often perceiving familiar shapes and patterns even in random objects and outlines. This phenomenon is known as pareidolia, and it's what makes us see fluffy sheep in clouds, funny faces in bumpy vegetables, or even the "man" in the moon. Psychologists have conducted studies revealing that when our eyes detect something, our brains scan it for recognizable grids and patterns, allowing us to quickly classify our perceptions and react. In evolutionary biology, this is seen as a life-saving primal instinct. The rule of thumb is simple: if something looks like a wolf, it could be a wolf, so the sooner you start, the better.

As we endeavor to comprehend the unfamiliar, our brains delight in being able to attribute a strong emotion to whatever structures it recognizes. Thus, even the most exquisite orchid blooms become comprehensible to us as soon as we detect in them a resemblance to faces or animals. When a peculiar new orchid species is discovered, we promptly find a catchy name, thanks to the gift of pareidolia. This explains why our world is filled with the grimaces of demons or white doves or what have you – and they are all flowers.

Orchideen mit Antlitz

DAS PHÄNOMEN DER PAREIDOLIE

WARUM WIR IN VIELEN DINGEN GESICHTER SEHEN: EINE PSYCHOLOGISCHE ERKLÄRUNG

Der Mensch liebt Gewohnheiten, und so neigen wir alle dazu, aus der Betrachtung zufälliger Formen und Umrisse bekannte Dinge zu abstrahieren. Pareidolie ist der Name des Phänomens, das uns in Wolken flauschige Schäfchen, in knubbeligem Gemüse drollige Gesichter oder auch den berühmten „Mann im Mond" sehen lässt. Psychologen haben es ergründet: Was die Augen sehen, scannt das Gehirn auf bekannte Raster und Muster ab, damit wir unsere Wahrnehmungen schnell einordnen und reagieren können. Evolutionsbiologisch ist das gewissermaßen ein lebensrettender Urinstinkt, nach dem Motto: „Sieht aus wie ein Wolf, könnte ein Wolf sein – besser schnell weg hier!"

Im Bestreben, Vertrautes in Unbekanntem zu entdecken, ist unser Gehirn geradezu entzückt, wenn es der erkannten Struktur gleich eine starke Emotion zuschreiben kann. So werden auch wundersam geformte Orchideenblüten für uns fassbar, sobald wir in ihnen eine Ähnlichkeit mit Gesichtern oder Tieren feststellen. Wenn nun eine bizarr geformte neue Orchideenart entdeckt wird, finden wir dank Pareidolie-Begabung schnell einen einprägsamen Namen – und willkommen in der Welt, ihr Teufelsfratzen und Friedenstauben in Blütenform!

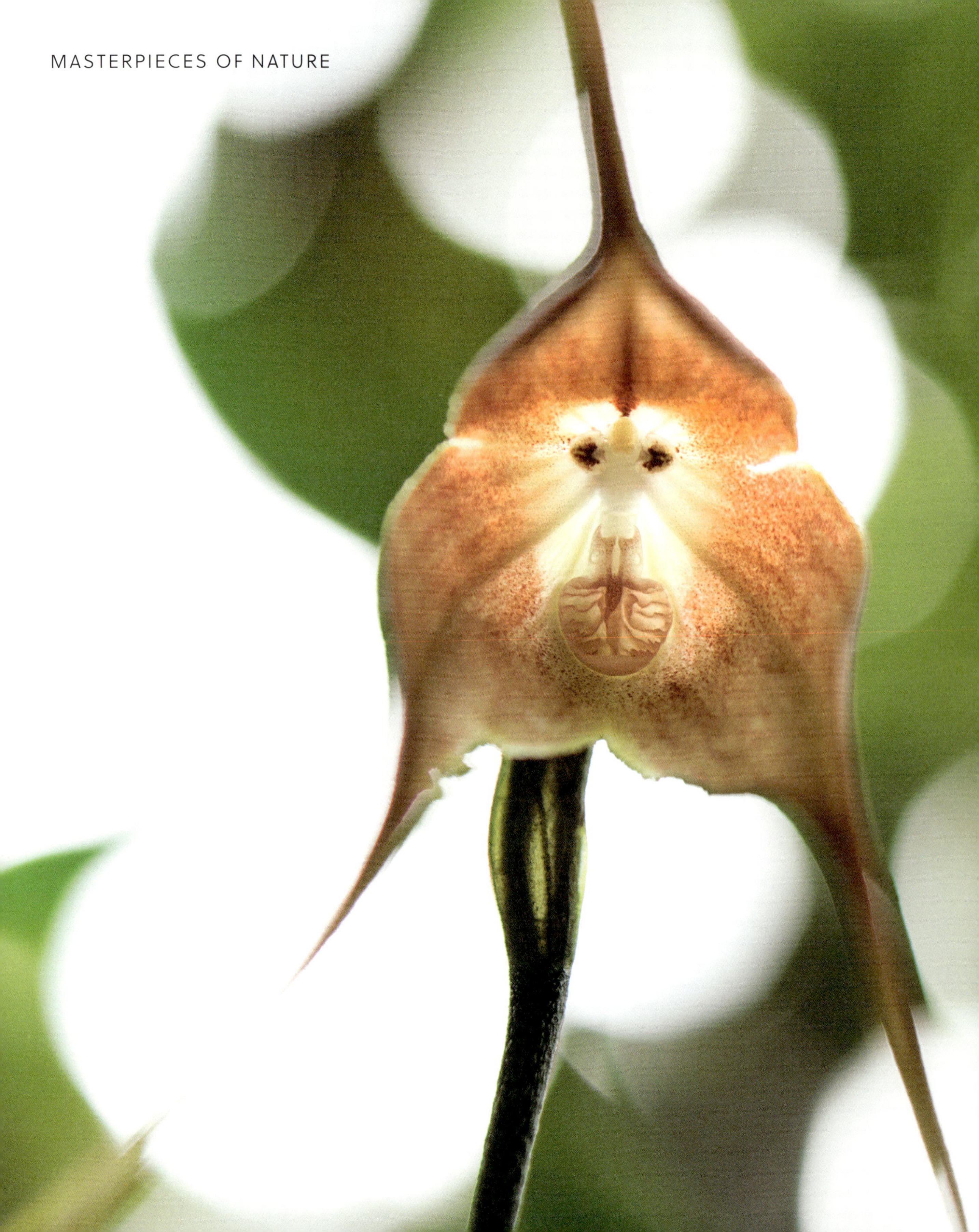

Dracula simia

MONKEY ORCHID

To humans, the flower of *Dracula simia* resembles the face of a monkey, but fruit flies perceive the mouth of this monkey as a fungus on which they can lay their eggs. They end up fertilizing the orchid.

AFFENGESICHT-ORCHIDEE

Für Menschen ähnelt die Blüte der *Dracula simia* einem Affengesicht, Fruchtfliegen nehmen das „Affenmaul" hingegen als Pilz wahr, auf dem sie Eier ablegen können, und befruchten so die Orchidee.

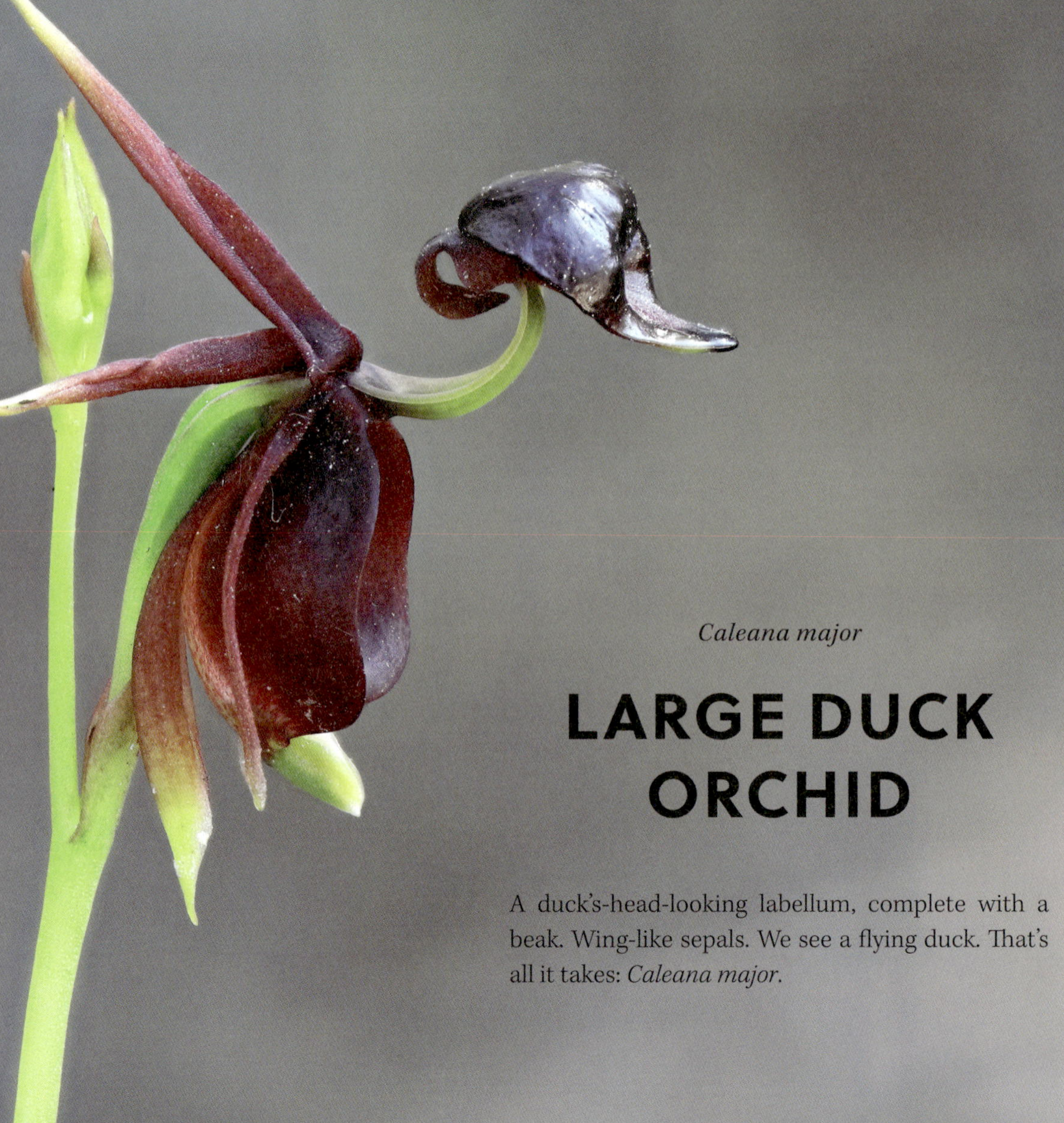

Caleana major

LARGE DUCK ORCHID

A duck's-head-looking labellum, complete with a beak. Wing-like sepals. We see a flying duck. That's all it takes: *Caleana major*.

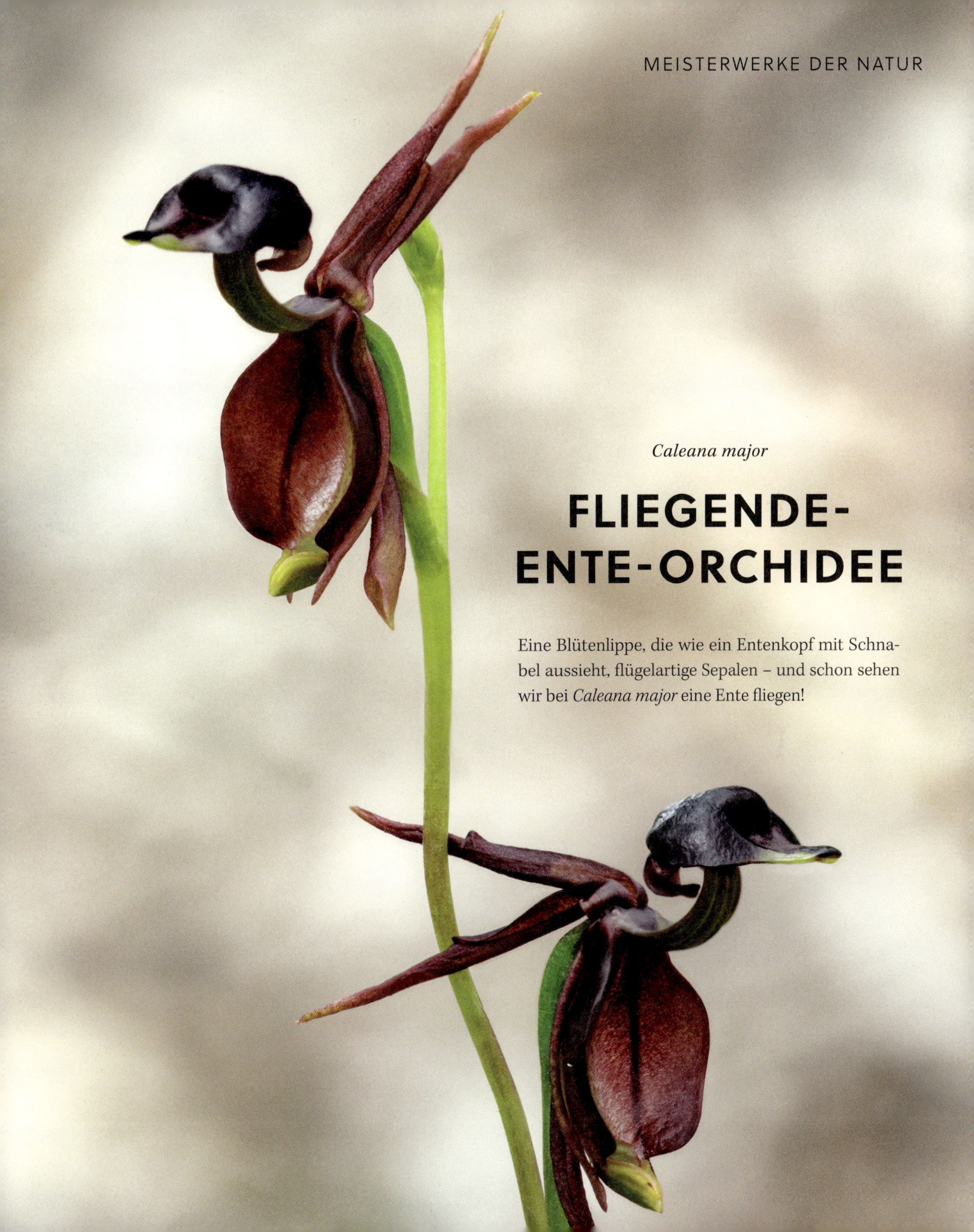

Caleana major

FLIEGENDE-ENTE-ORCHIDEE

Eine Blütenlippe, die wie ein Entenkopf mit Schnabel aussieht, flügelartige Sepalen – und schon sehen wir bei *Caleana major* eine Ente fliegen!

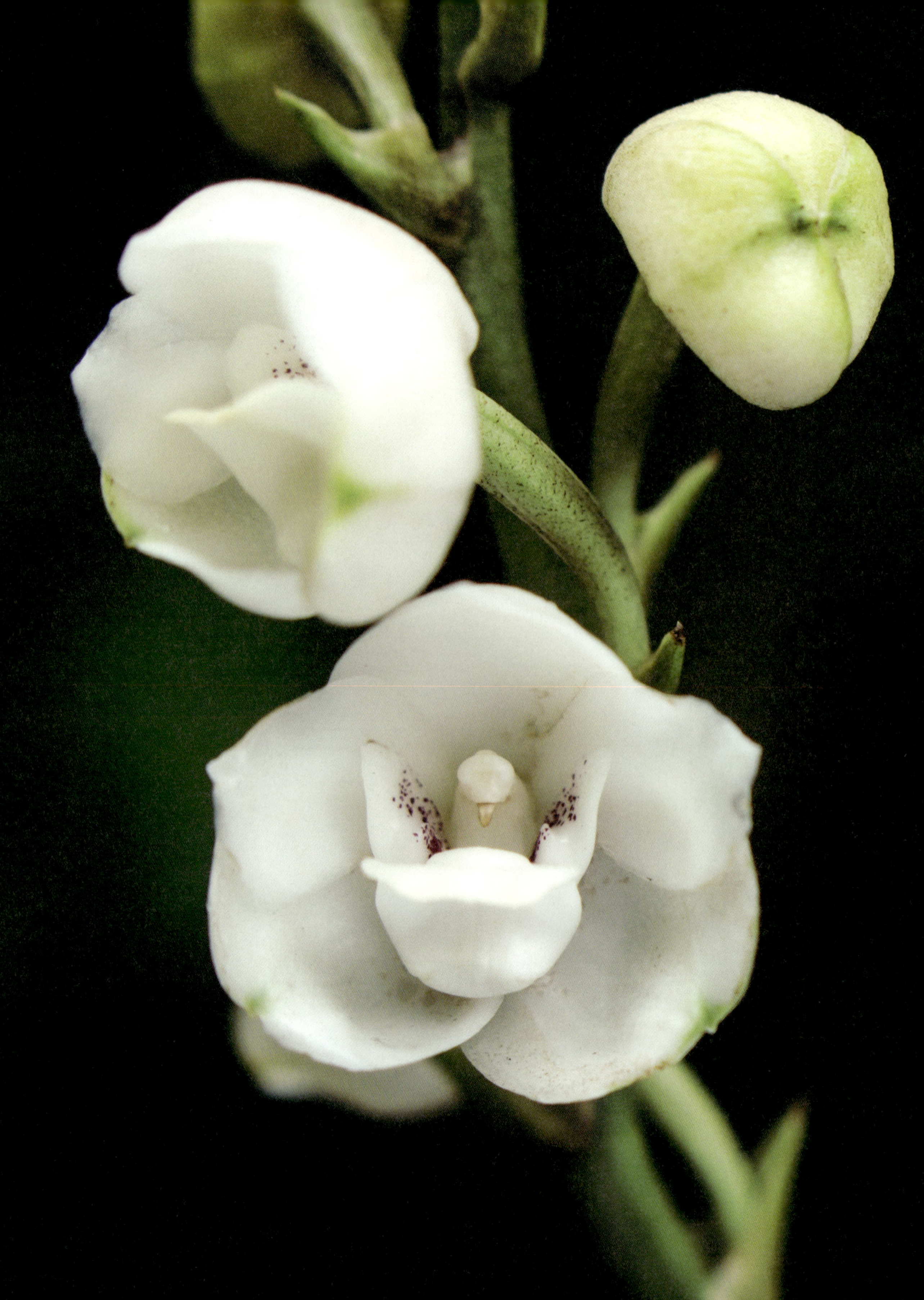

Peristeria elata

DOVE ORCHID

As if a tiny dove had settled in the center of the flower! For the same reason, *Peristeria elata* is also called the Holy Ghost orchid. It is the national flower of Panama.

TAUBEN-ORCHIDEE

Als hätte sich ein winziges Täubchen in der Blüte niedergelassen – die *Peristeria elata* wird deshalb auch „Blume des Heiligen Geistes" genannt und gilt als Nationalblume Panamas.

Pecteilis radiata

WHITE EGRET ORCHID

It looks like a great egret has spread its wings in flight. *Pecteilis radiata* is enchanting with its delicately "feathered" petals.

WEISSE VOGELBLUME

Als hätte ein Silberreiher seine Schwingen im Flug gespreizt: Die *Pecteilis radiata* verzaubert mit ihren fein gefiederten Blütenblättern.

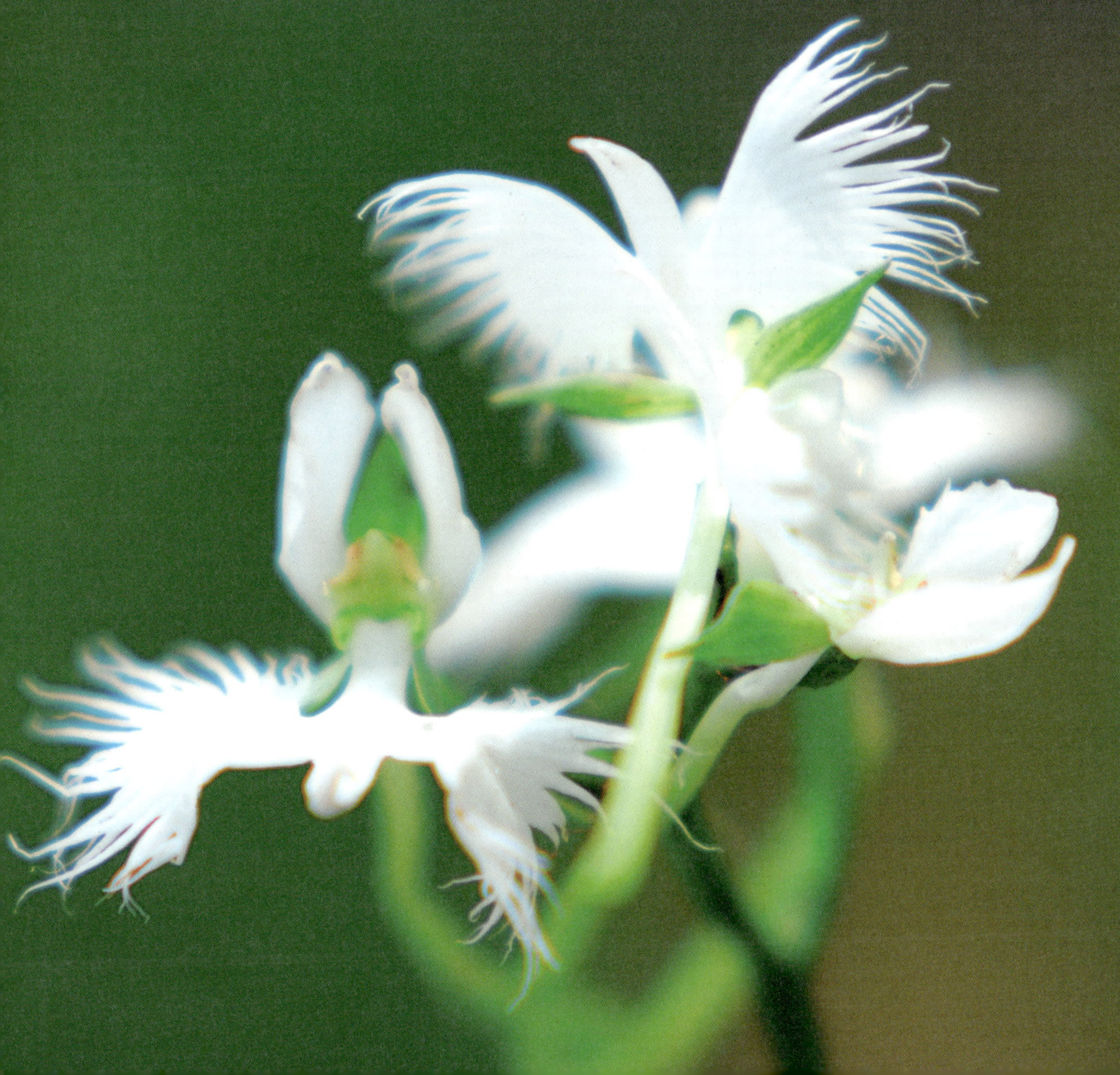

Prosthechea cochleata

CLAMSHELL ORCHID

If you consider the entirety of the flower of *Prosthechea cochleata*, it looks like an octopus. By itself, the labellum alone resembles a clam or even a cockle. It is the national flower of Belize.

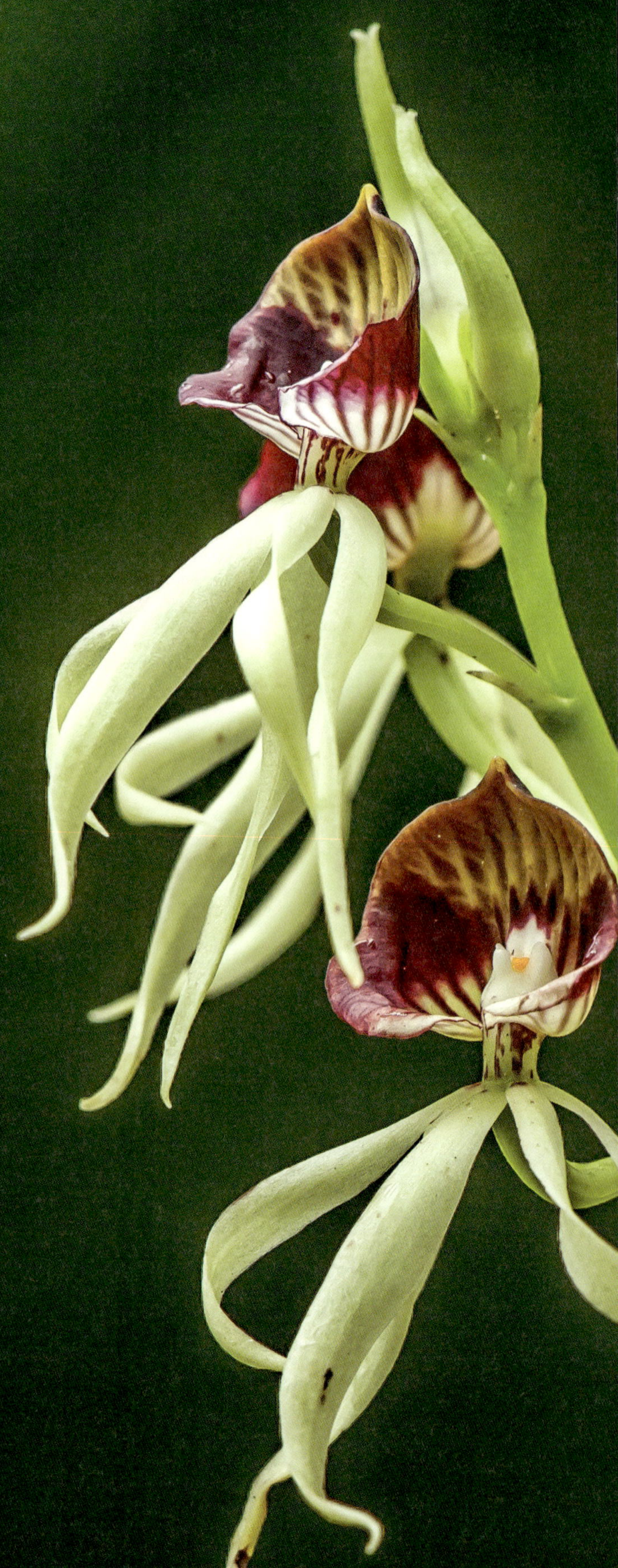

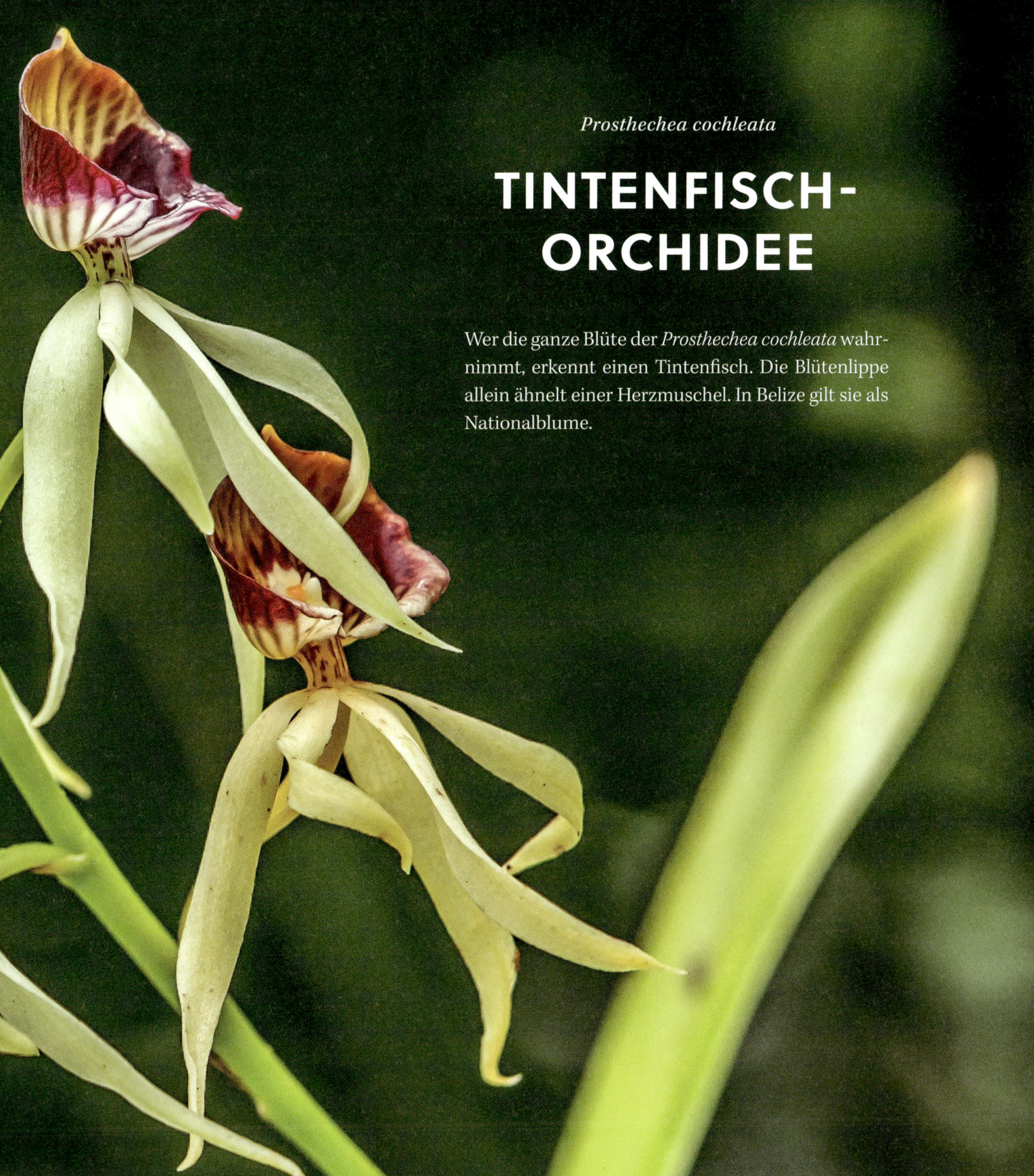

Prosthechea cochleata

TINTENFISCH-ORCHIDEE

Wer die ganze Blüte der *Prosthechea cochleata* wahrnimmt, erkennt einen Tintenfisch. Die Blütenlippe allein ähnelt einer Herzmuschel. In Belize gilt sie als Nationalblume.

Ophrys apifera

BEE ORCHID

The fine lateral hairs on the flower are probably what make *Ophrys apifera* seem to resemble a bee.

BIENEN-RAGWURZ

Es sind wohl die feinen Haare seitlich an der Blüte, die bei *Ophrys Apifera* die Ähnlichkeit mit einer Biene ausmachen.

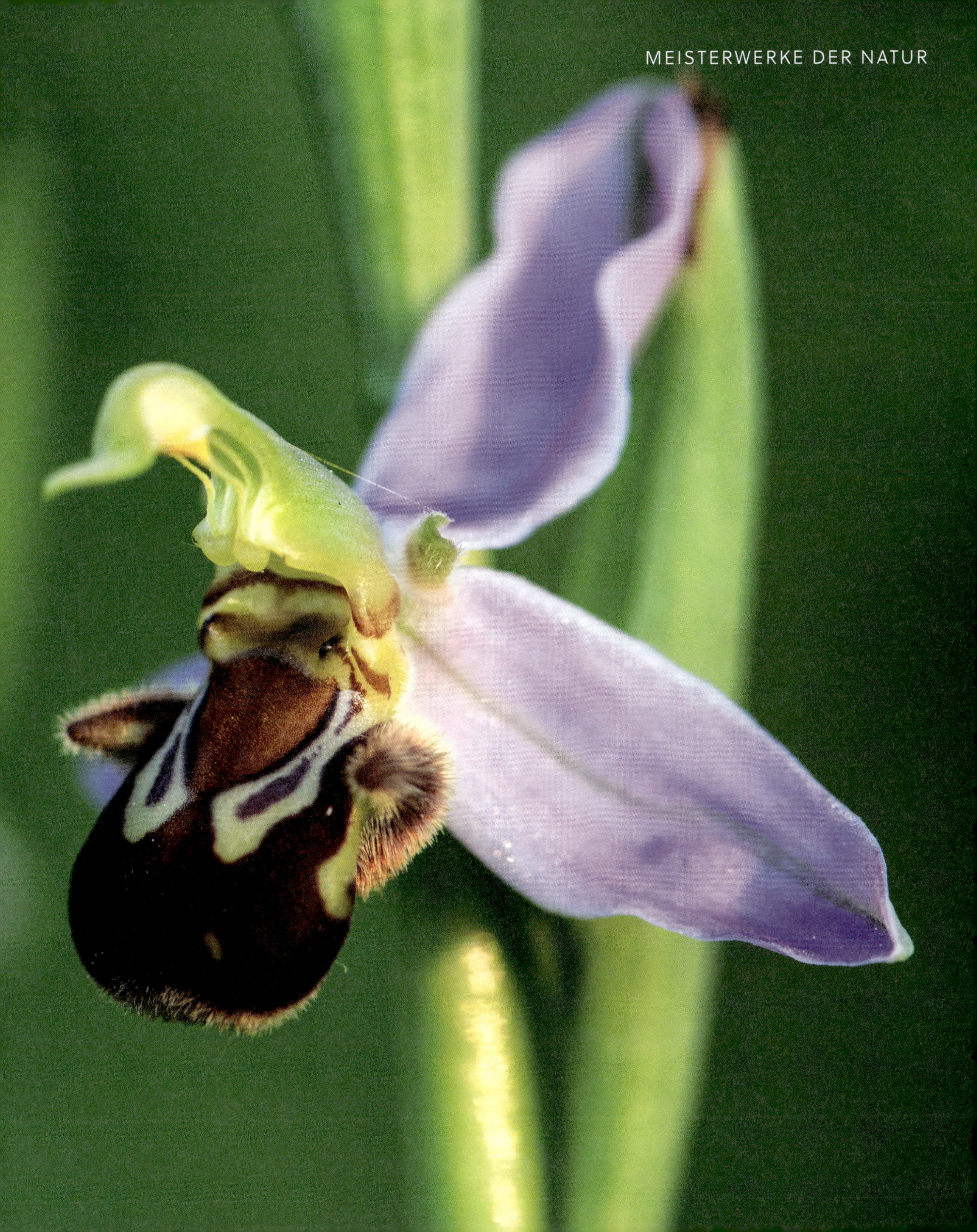

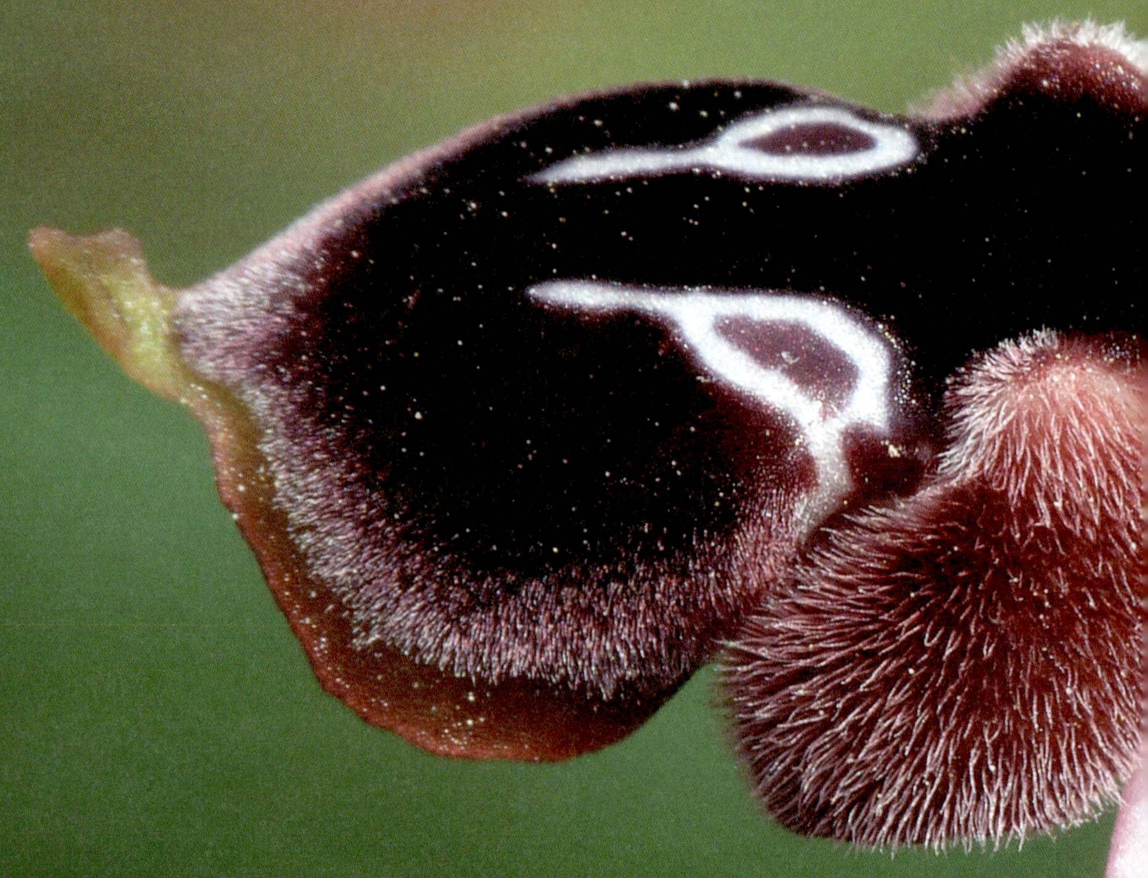

Ophrys reinholdii

GOAT ORCHID

There are varying opinions about *Ophrys reinholdii*; some people see the flower as a hornet, while others perceive it as the horned head of a goat, which explains its English name, the goat orchid.

Ophrys reinholdii

REINHOLDS RAGWURZ

Bei *Ophrys reinholdii* gehen die Meinungen ausein-
ander: Manch einer nimmt die Blüte als Hornisse
wahr, andere als gehörnten Kopf einer Ziege – daher
auch der englische Name *Goat Orchid*.

Ophrys insectifera

FLY ORCHID

The petals look like little fake wings. Together with the scent of *Ophrys insectifera*, the flower makes such a realistic impression that male flies try to mate with it.

FLIEGEN-RAGWURZ

Die Blütenblätter täuschen die Flügel vor – zusammen mit dem Duftstoff der *Ophrys insectifera* so realistisch, dass Fliegenmännchen versuchen, die Blüte zu begatten.

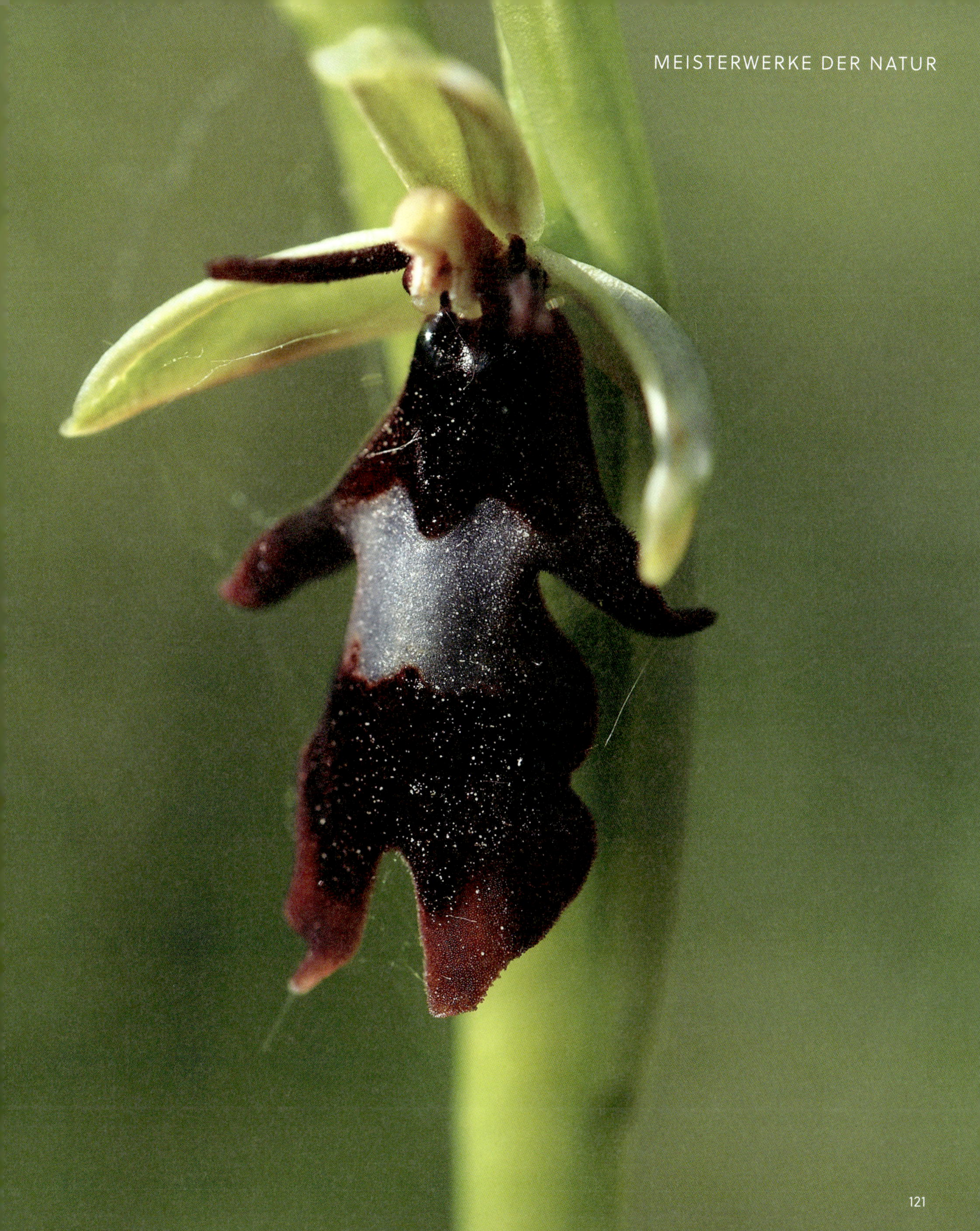

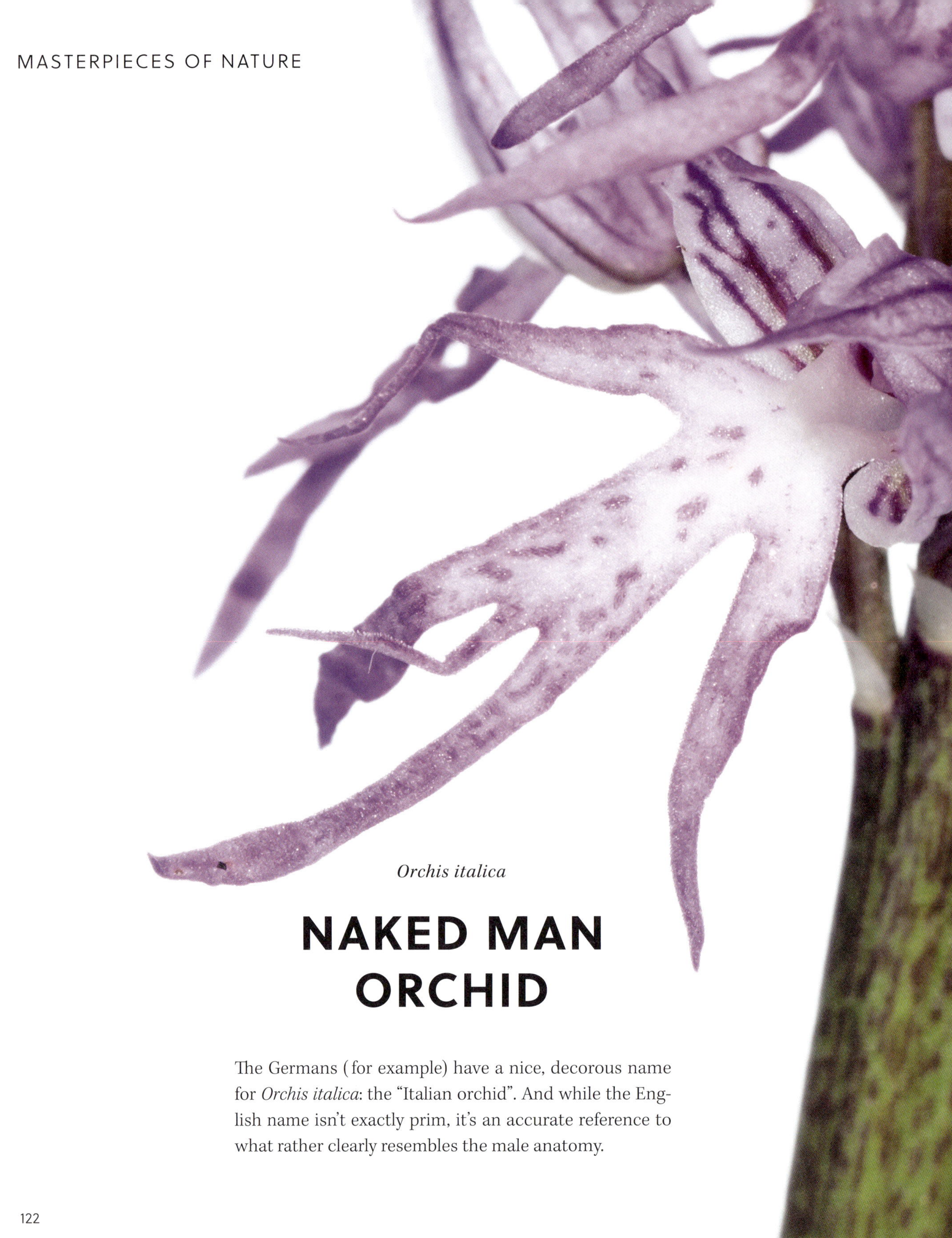

Orchis italica

NAKED MAN ORCHID

The Germans (for example) have a nice, decorous name for *Orchis italica*: the "Italian orchid". And while the English name isn't exactly prim, it's an accurate reference to what rather clearly resembles the male anatomy.

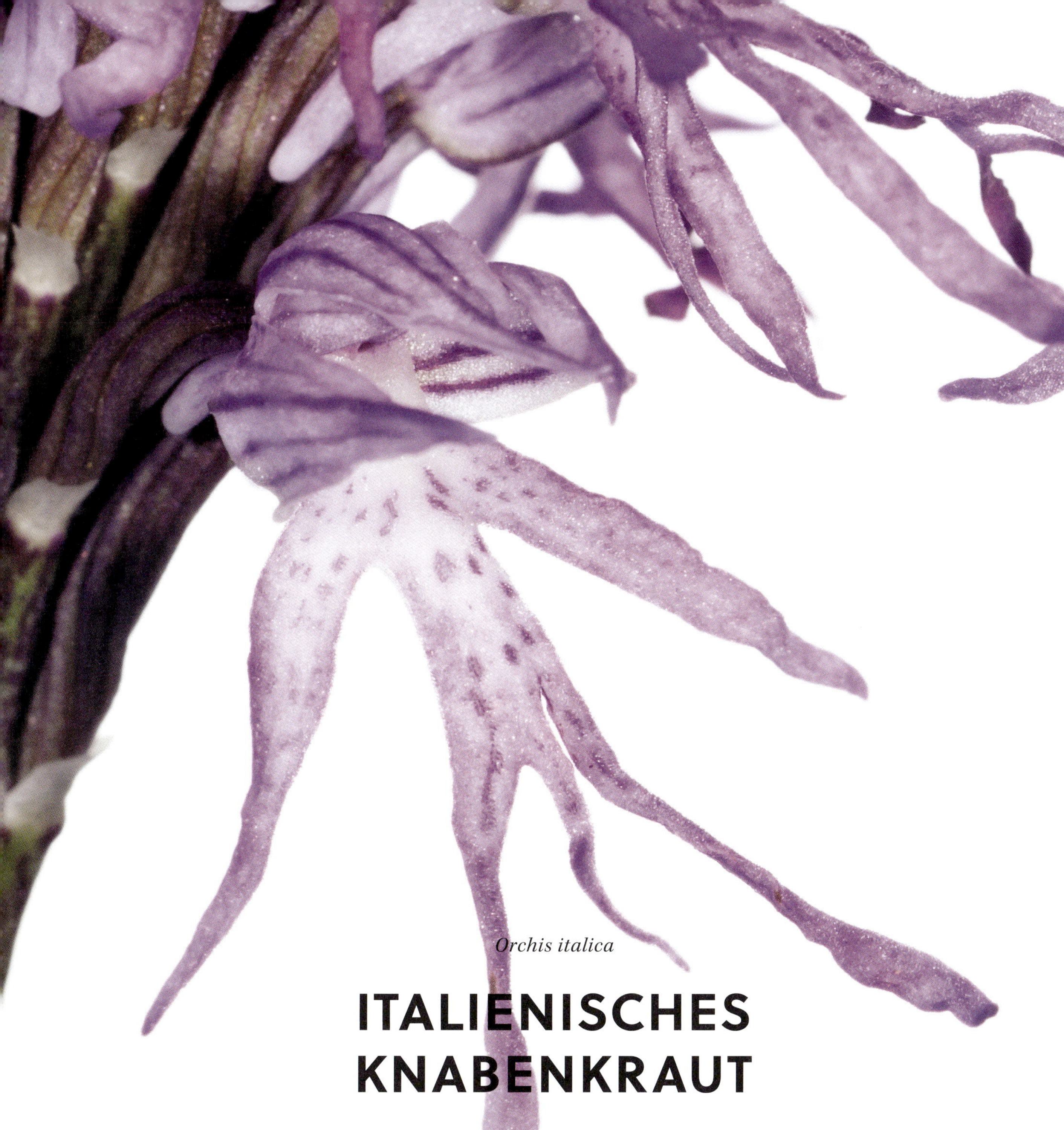

Orchis italica

ITALIENISCHES KNABENKRAUT

Orchis italica ist im englischsprachigen Raum auch als Nackter-Mann-Orchidee bekannt – ein Blick auf die Blütenform zeigt deren Ähnlichkeit mit der männlichen Anatomie.

Grammatophyllum speciosum

BIG, GIANT, TIGER ORCHID

Grammatophyllum speciosum is regarded as the largest orchid species in the world. Its sturdy plant stem can reach a length of about ten feet, while its inflorescence can grow to more than six feet long and bear as many as forty flowers. These flowers are of an orange-brown hue, reminiscent of the fur of a tiger.

GROSS, GRÖSSER, TIGER-ORCHIDEE

Die *Grammatophyllum speciosum* gilt als größte Orchideenart der Welt: Ihre kräftigen Sprossachsen werden bis zu drei Meter lang, ihre Blütenstände tragen an einer Länge von rund zwei Metern bis zu 40 Blüten, die orange-braun gefärbt sind wie das Fell eines Tigers.

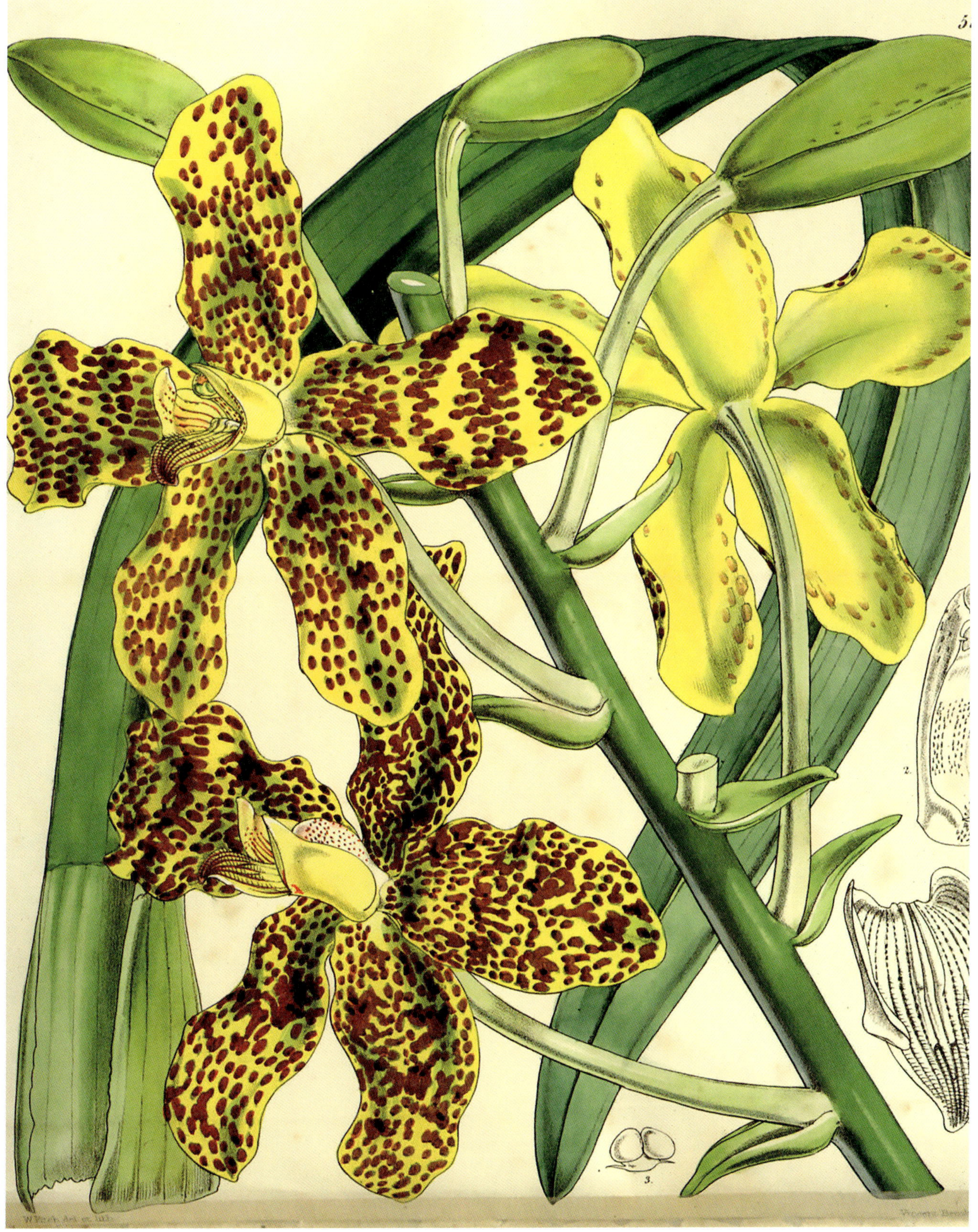

W. Fitch del. et lith.
Vincent Brooks
5.
2.
3.

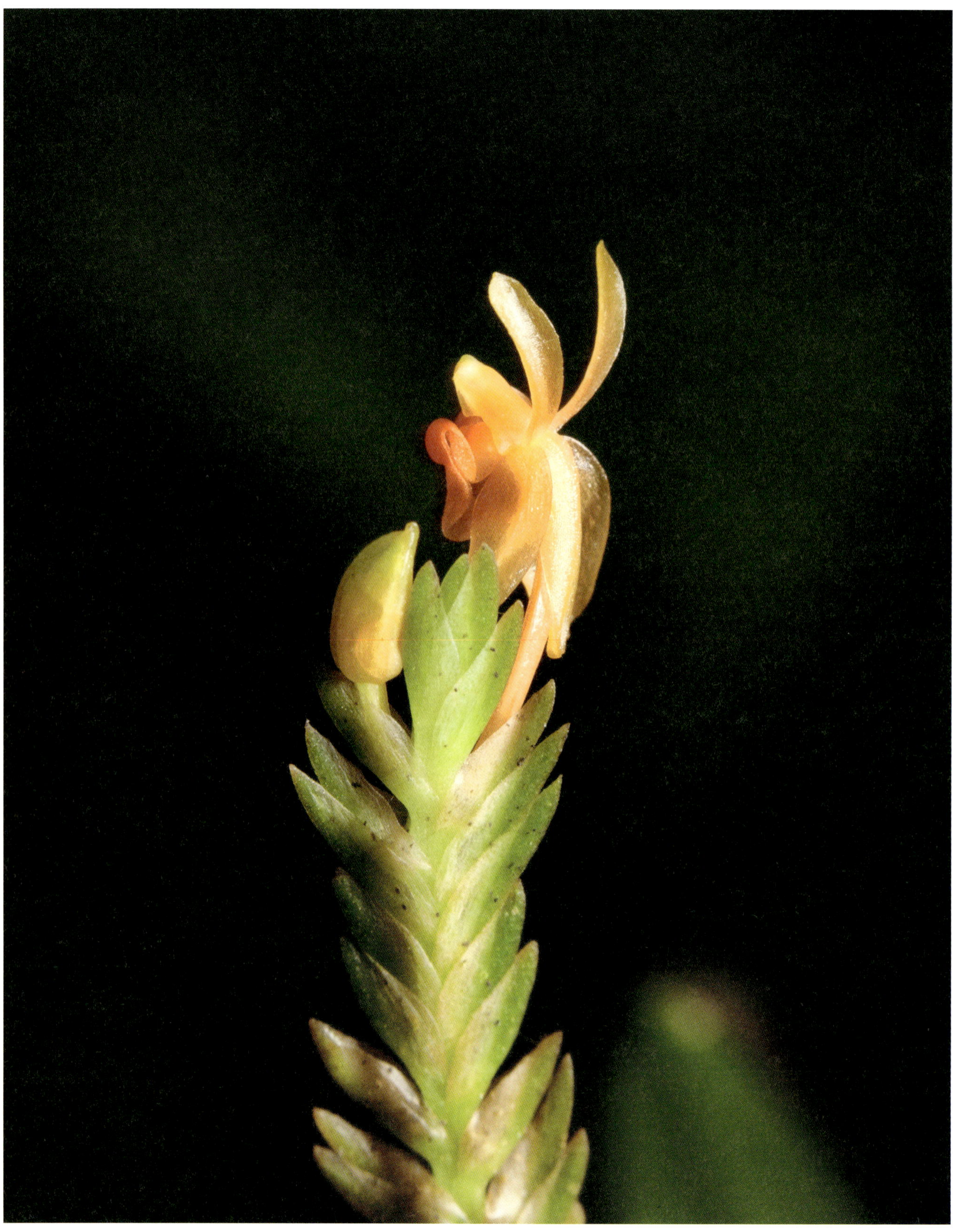

Campylocentrum insulare

THE LILLIPUTIAN ORCHID

Campylocentrum insulare is now recognized as the world's smallest orchid, and it was discovered in Santa Catarina, Brazil by sheer coincidence. In 2015, while examining this tiny object under the microscope, researchers initially mistook its 0.5-millimeter flower for a fungus. Prior to this discovery, the record for the smallest flower was held by *Lepanthes oscarrodrigoi*, which was six times larger.

DIE „LILLIPUT"-ORCHIDEE

Die heute als kleinste Orchidee der Welt geltende *Campylocentrum insulare* verdankt ihre Entdeckung im brasilianischen Santa Catarina einem Zufall: Unter das Mikroskop der Forscher kam sie im Jahr 2015, weil man die 0,5 Millimeter kleine Orchideenblüte für einen Pilz hielt! Bis dahin wurde der Rekord der Blütenwinzlinge von *Lepanthes oscarrodrigoi* mit 3 Millimetern gehalten

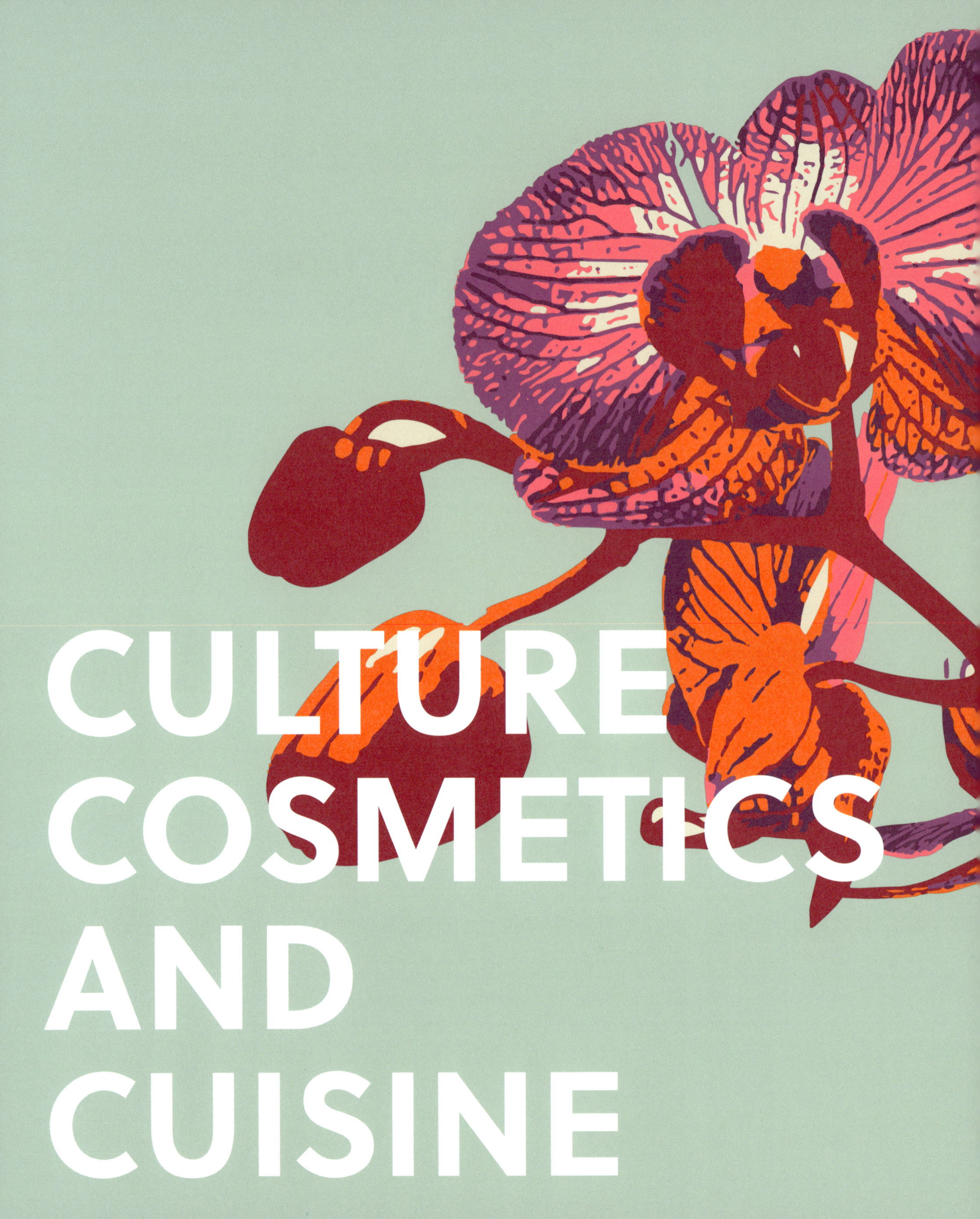

CULTURE COSMETICS AND CUISINE

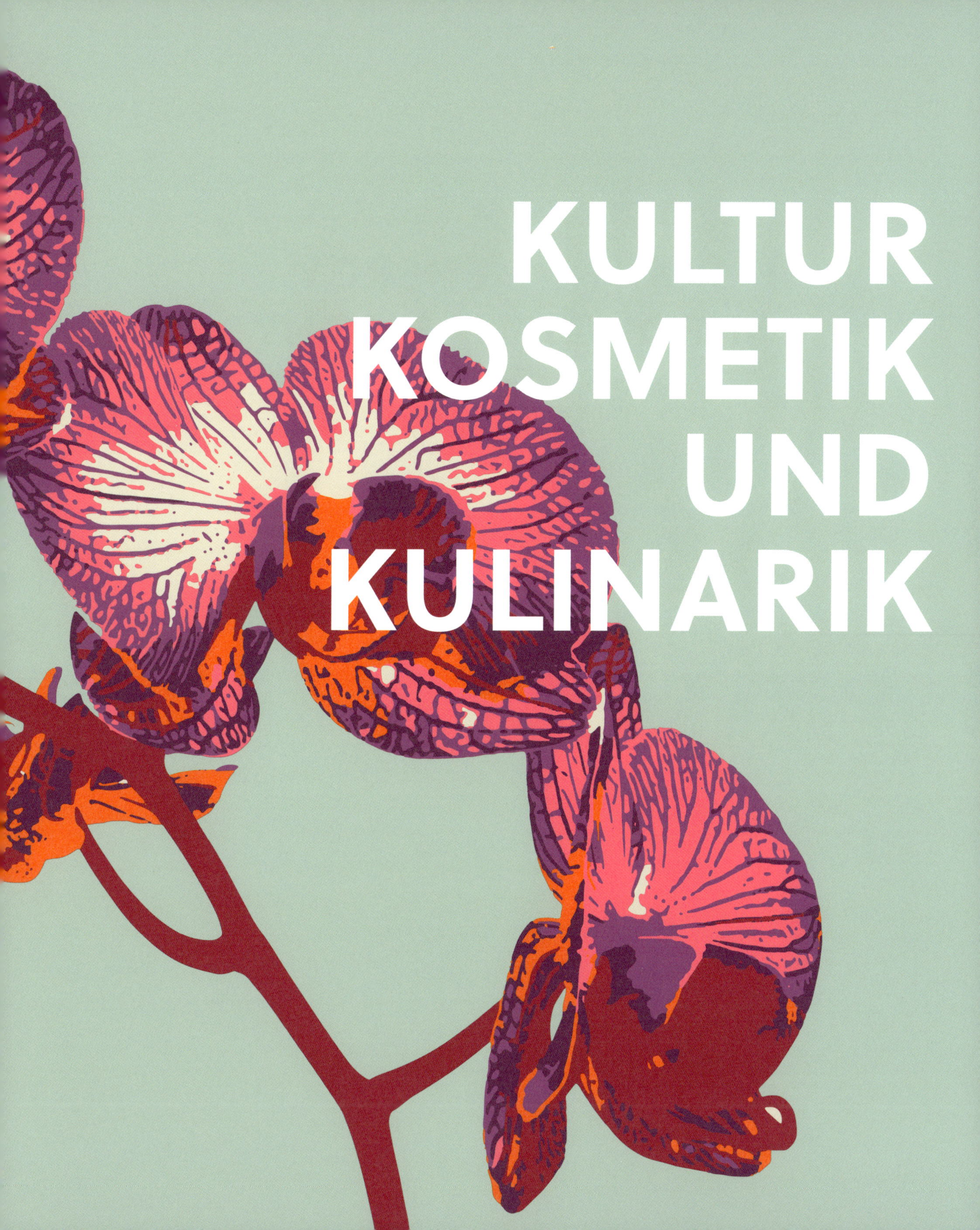
KULTUR
KOSMETIK
UND
KULINARIK

Feng Shui

POSITIVE ENERGY THROUGH HARMONY

IN ASIA, THE ORCHID IS REGARDED AS A SYMBOL OF REFINEMENT, BALANCE, LOVE AND FERTILITY

In Chinese art, the orchid represents the arrival of spring and, together with bamboo, plum blossom, and the chrysanthemum, forms part of the *sì jūnzǐ* quartet of flowers – the "Four Gentlemen". And so it is that the orchid holds a permanent place in Feng Shui, the Chinese theory of harmony. It is often used to promote the energies of love, partnership, and fertility, but also as a plant symbol of patience, inner balance, and order.

NEXT PAGE | Besides boding well in terms of wealth and success, purple orchids are said to bring home great joy.

Feng-Shui

POSITIVE ENERGIE DURCH HARMONIE

IN ASIEN GILT DIE ORCHIDEE ALS SYMBOL FÜR ELEGANZ, HARMONIE, LIEBE UND FRUCHTBARKEIT

In der chinesischen Kunst symbolisiert die Orchidee den Frühling und zählt neben Bambus, Pflaumenblüte und Chrysantheme zum Quartett der *Sì Jūnzǐ*, was so viel wie „Die vier Edlen" bedeutet. So hat die Orchidee auch ihren festen Platz in der chinesischen Harmonielehre des Feng-Shui. Dort wird sie gerne als Energiepflanze für Liebe, Partnerschaft und Fruchtbarkeit verwendet, gilt aber auch als Symbol für Geduld, innere Harmonie und Ordnung.

NÄCHSTE SEITE | Lilafarbene Orchideen sollen besonders viel Glück ins Haus bringen. Außerdem stehen sie für Erfolg und Wohlstand.

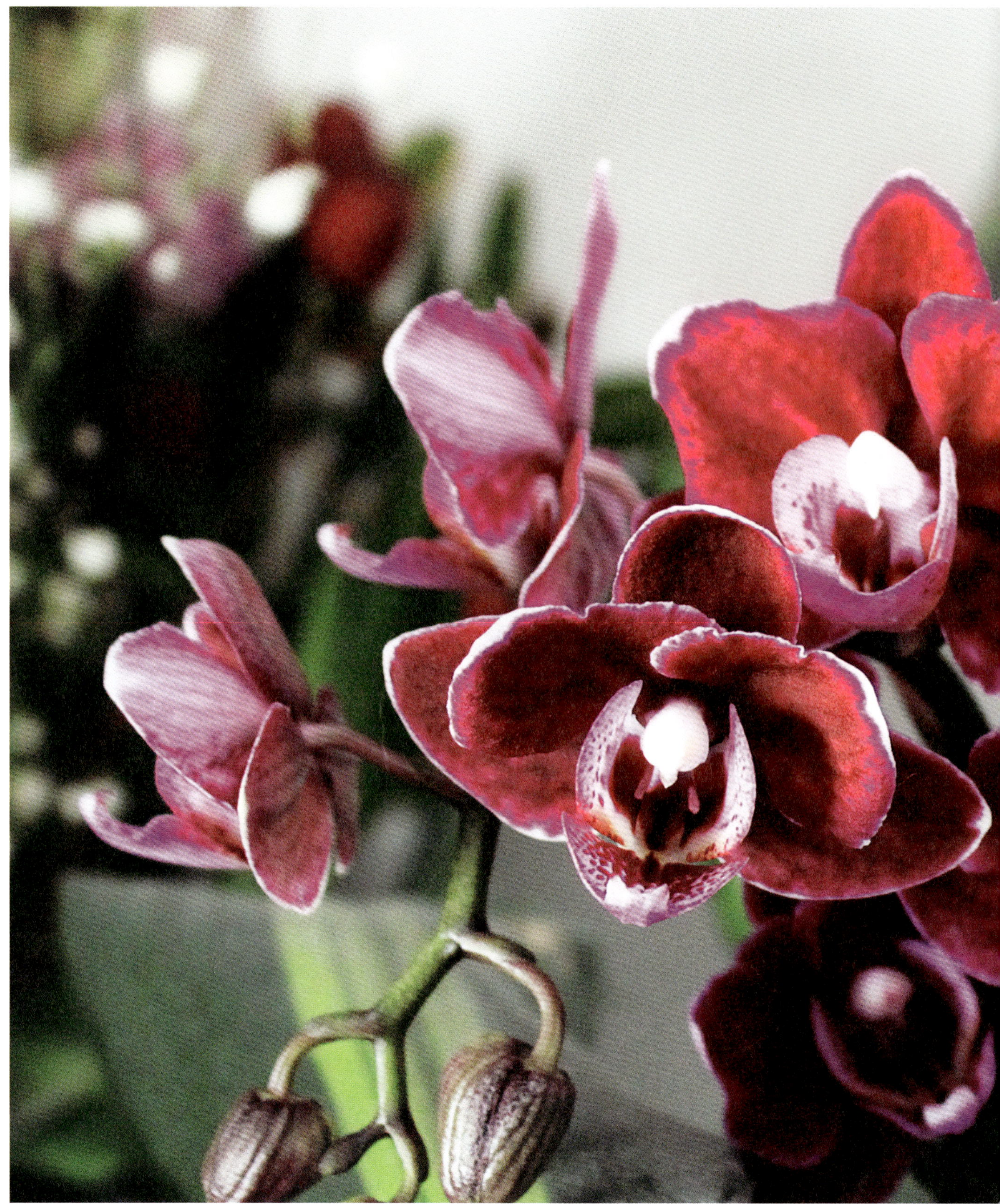

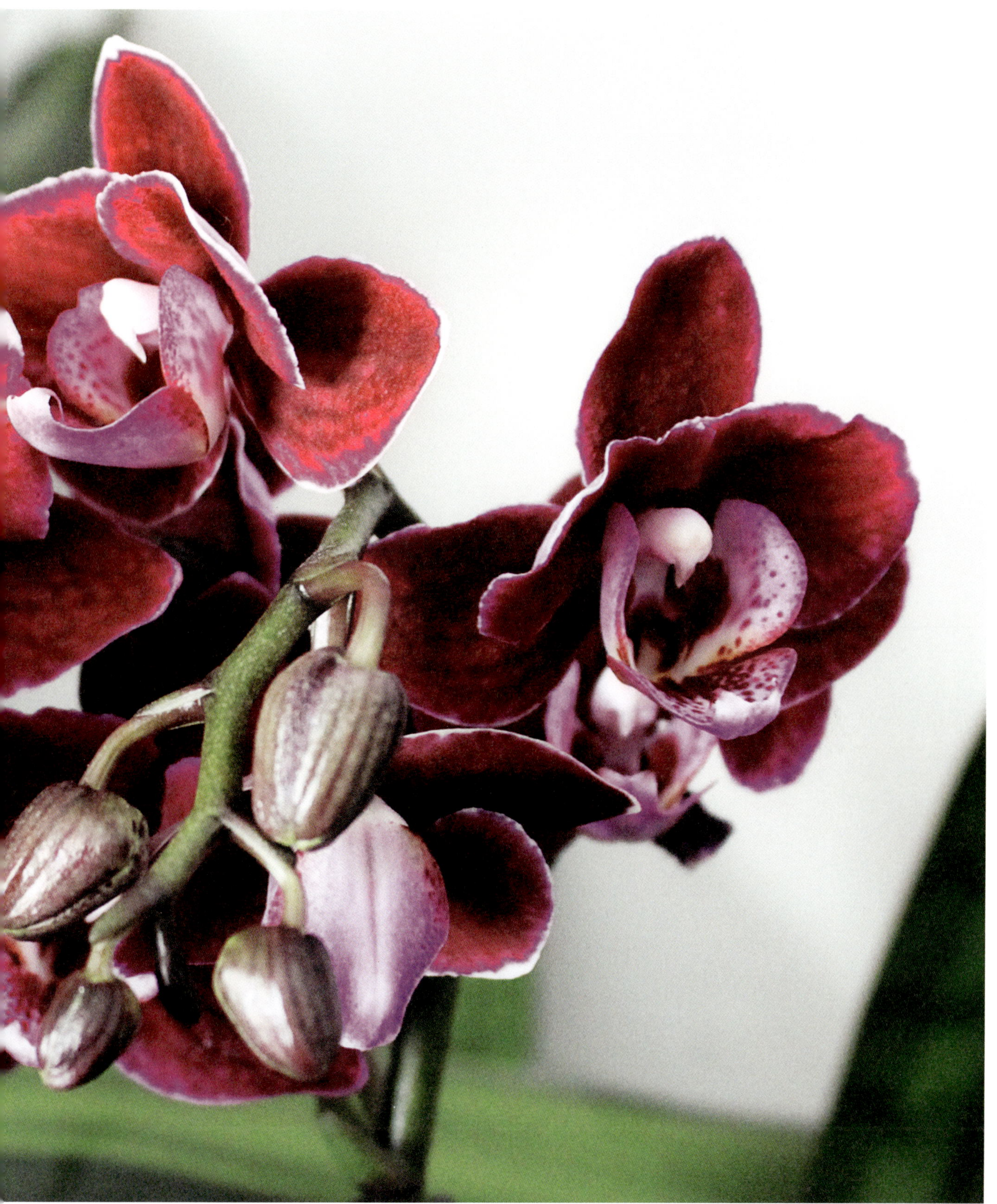

Dendrobium chrysotoxum

THE ORCHIDS OF MR. MARGRAF

ARE ORCHID FLOWERS THE SECRET FOUNTAIN OF YOUTH? THE COSMETICS INDUSTRY THINKS SO

Located within a nature reserve in the Chinese province of Yunnan, near the borders with Myanmar and Laos, lies a veritable orchid paradise. This wondrous place exists in its current form thanks to the efforts of German biologist Josef Margraf and a French cosmetics manufacturer. The misty rainforests here offer ideal conditions for moisture-loving orchids. To preserve endangered *Vanda* and *Dendrobium* species, Margraf, with support from the cosmetics company, initiated a rainforest reforestation project. Orchids that grow in the rainforest are considered more vital and valuable than those cultivated in greenhouses, making the rainforest an interesting location for cosmetics research. This creates a win-win situation for both nature and business.

During the flowering period, which lasts for only three weeks, the orchid splendor in Yunnan is painstakingly hand-harvested from high treetops. Of particular interest to the cosmetics industry are the "golden-bow" *Dendrobium chrysotoxum*, *Vanda coerulea* and *Vanda teres*. These orchids are not only prized for their fragrances but also for the substances that are responsible for the extreme longevity of their flowers. The beauty industry is therefore keen to extract these substances to create an "elixir of eternal youth" that can slow down the aging of skin.

NEXT PAGE | Two of the coveted orchids *Dendrobium chrysotoxum* (left) and *Vanda coerulea* (right)

Dendrobium chrysotoxum

DIE ORCHIDEEN VON HERRN MARGRAF

STECKT IN ORCHIDEENBLÜTEN DAS GEHEIMNIS EWIGER JUGEND? DIE KOSMETIKBRANCHE GLAUBT DARAN …

In einem Naturreservat in der chinesischen Provinz Yunnan, nahe der Grenzen zu Myanmar und Laos, liegt ein Paradies der Orchideen. Dass es in dieser Form heute existiert, ist auch dem deutschen Biologen Josef Margraf und einem französischen Kosmetikhersteller zu verdanken. Die nebligen Regenwälder bieten feuchtigkeitsliebenden Orchideen ideale Bedingungen, und so initiierte Margraf, unterstützt durch den Kosmetikkonzern, dort ein Regenwald-Aufforstungsprojekt, um gefährdete *Vanda*- und *Dendrobium*-Arten zu erhalten. Die im Regenwald wachsenden Orchideen sollen vitaler und wertvoller sein als Gewächshauszüchtungen, weshalb die Kultivierung im Regenwald für die Kosmetikforschung interessant ist – eine Win-win-Situation für Natur und Wirtschaft.

In Yunnan wird die Orchideenpracht deshalb während der nur dreiwöchigen Blütezeit aufwendig von Hand in den hohen Baumkronen geerntet. Interessant für die Kosmetikbranche sind die Goldorchidee *Dendrobium chrysotoxum* sowie *Vanda coerulea* und *Vanda teres* nicht nur aufgrund ihres Duftes: Das Bestreben der Schönheitsindustrie ist es, die Stoffe zu extrahieren, die für die extreme Langlebigkeit von Orchideenblüten verantwortlich sind. So wird aus den Orchideen ein „Elixier der ewigen Jugend" gewonnen, das die Hautalterung aufhalten soll.

NÄCHSTE SEITE | Begehrt: Goldorchidee (*Dendrobium chrysotoxum, links*) und *Vanda coerulea* (rechts)

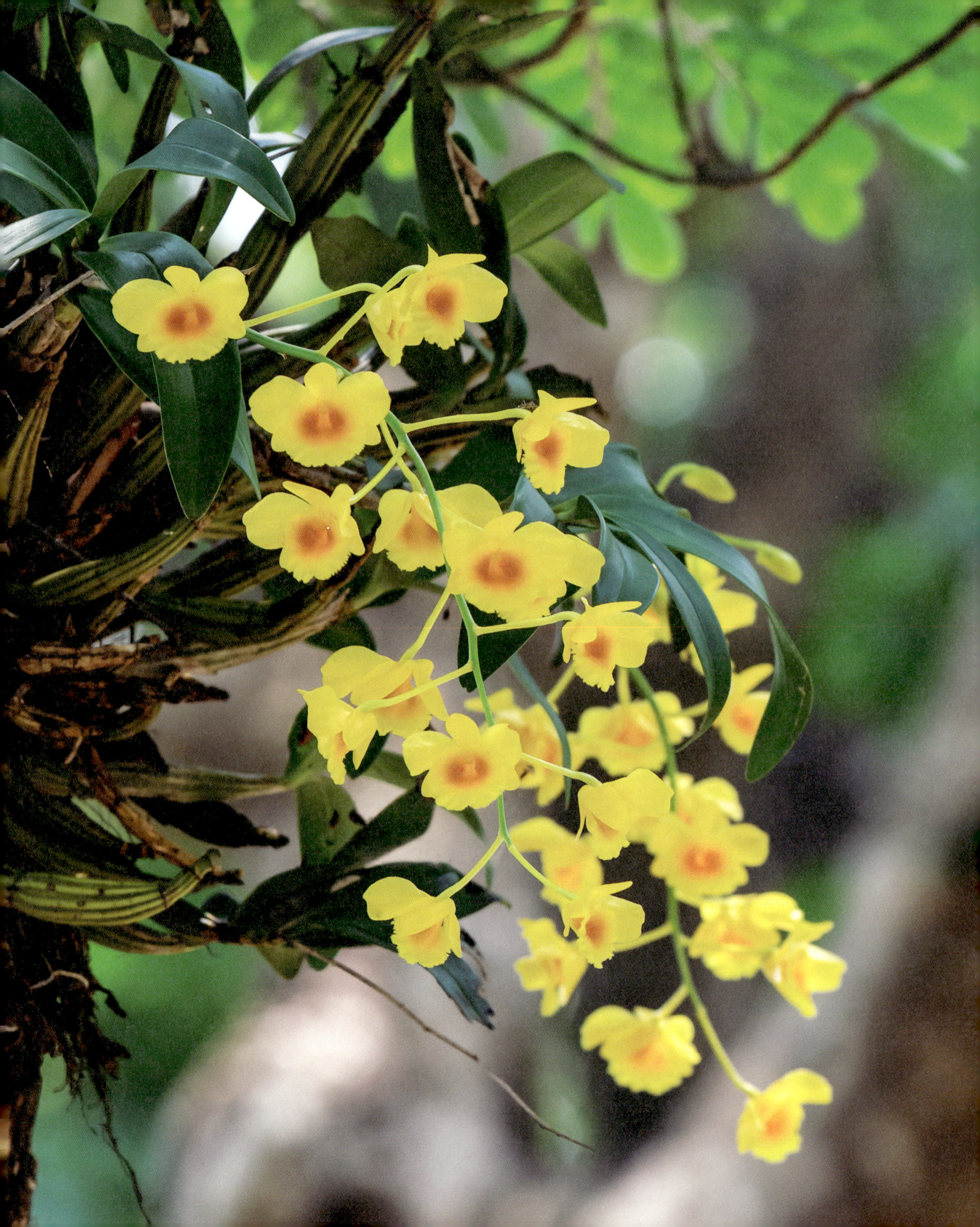

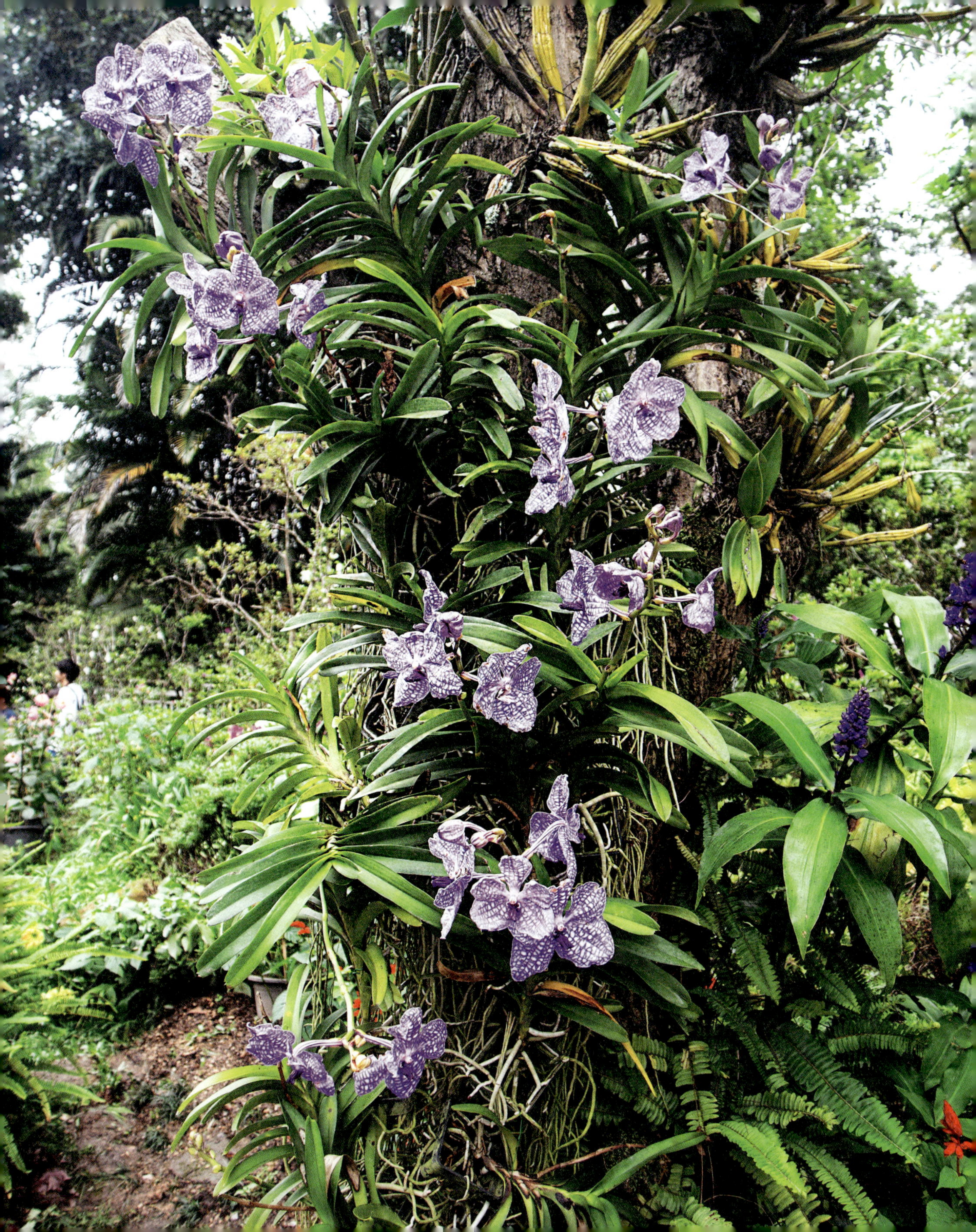

Vanilla planifolia

A GOURMET AROMA

VANILLA, ONE OF OUR MOST POPULAR SPICES, IS THE SWEET SECRET OF THE ORCHID WORLD

One of the most delectable flavors to enhance the culinary arts comes from the seed pods of one genus of orchid: *Vanilla planifolia*. Obtaining these aromatic pods from the vanilla orchid requires a combination of precision timing and patience. The green fruit can only be obtained if the vanilla is pollinated on the day of its short flowering. The pods are harvested about nine months later, and then blanched and slowly dried. It is only then that the fermented capsules, containing thousands of tiny seed globules that carry the incomparable vanilla flavor, are fit for consumption.

The vanilla orchid is native to Mexico and Central America, and since it can only be naturally pollinated by certain bees and hummingbirds, the "black flower of the Aztecs" remained confined to these regions for a long time. Vanilla reached Europe during the colonization and exploitation of the Aztecs in the early sixteenth century, via the conquistador Hernán Cortés, where it became a spice exclusively for monarchs and the wealthy. Large-scale cultivation of Bourbon vanilla did not begin until 1841. It became possible thanks to an enslaved twelve-year-old boy named Edmond Albus, who lived on La Réunion, formerly known as Île Bourbon. Albus discovered how to manually pollinate the blossom of the orchid after smugglers had imported the plant to the island.

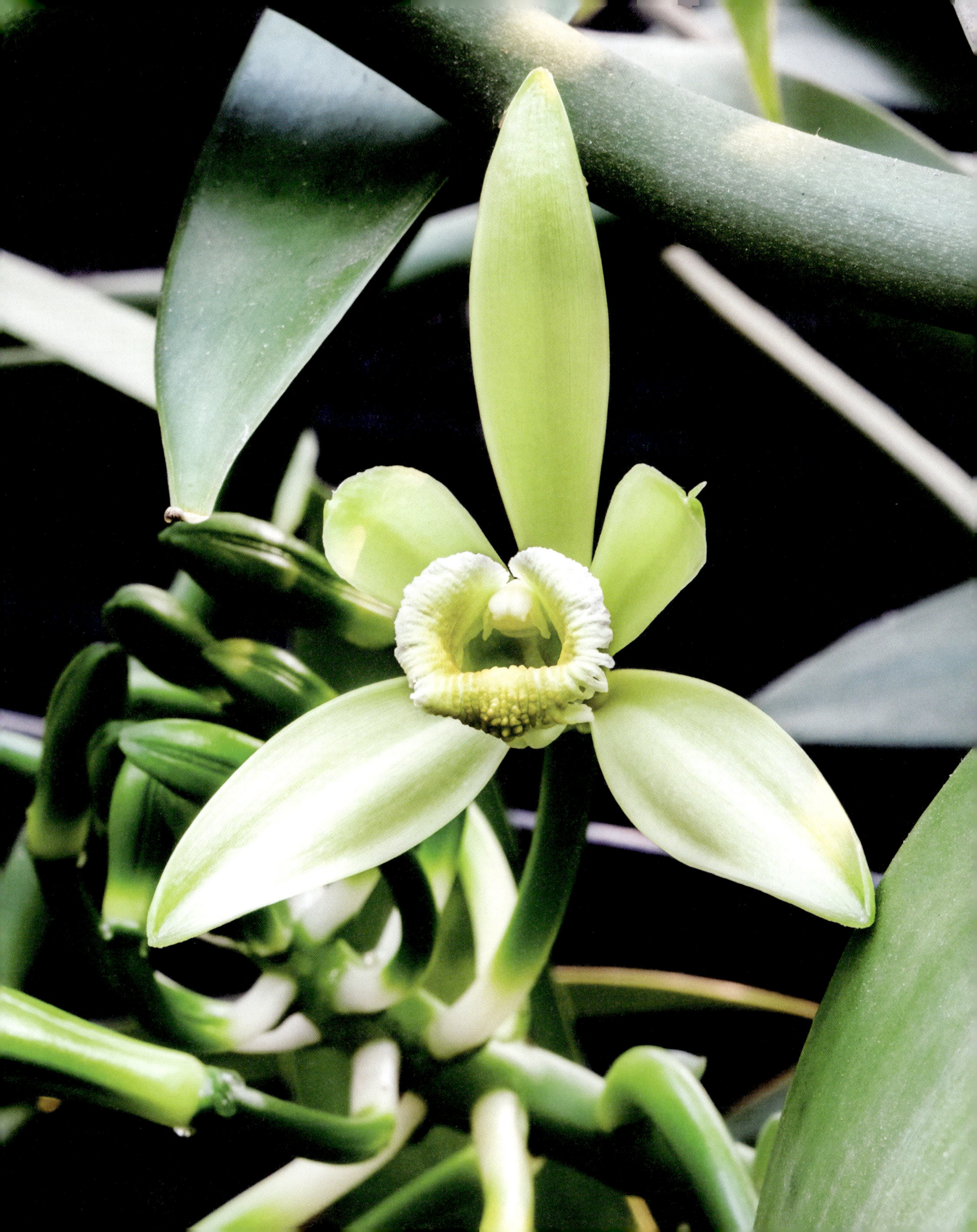

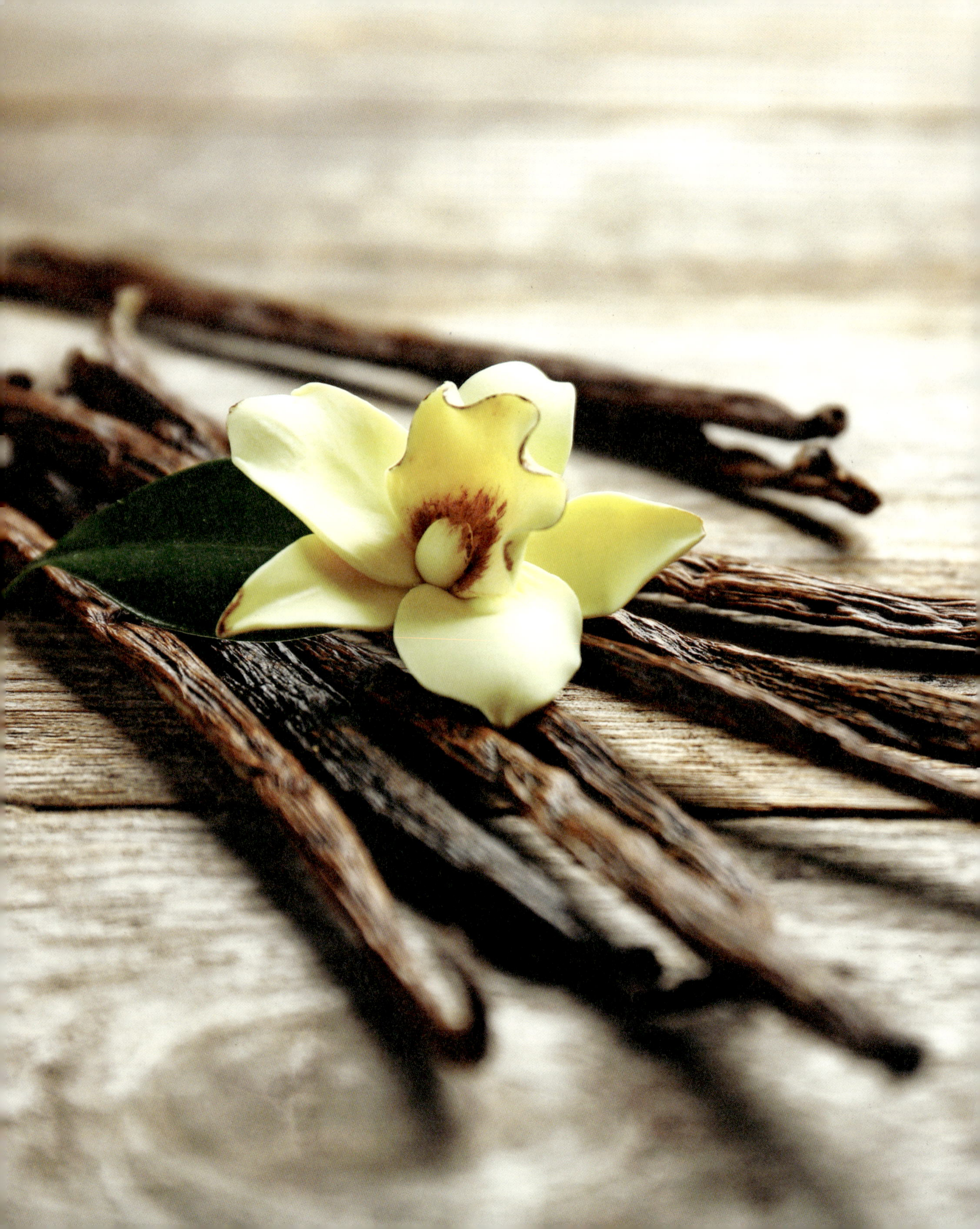

Vanilla planifolia

EIN AROMA FÜR GOURMETS

DAS SÜSSE GEHEIMNIS DER ORCHIDEENWELT IST EINES UNSERER BELIEBTESTEN GEWÜRZE: VANILLE

Eines der köstlichsten Aromen, die die Kunst des Kochens und Backens bereichern, stammt aus den Samenkapseln einer Orchideengattung: *Vanilla planifolia*. Um von der Vanilleorchidee die aromatischen Schoten ernten zu können, ist eine Mischung aus exaktem Timing und Geduld nötig: Nur, wenn die Vanille am Tag ihrer kurzen Blüte bestäubt wird, kann nach rund neun Monaten die grüne Frucht geerntet, blanchiert und langsam getrocknet werden. Erst dann sind die fermentierten Kapseln mit ihren tausenden von winzigen Samenkügelchen, die das unvergleichliche Vanillearoma tragen, reif zum Genuss.

Die Vanilleorchidee ist in Mexiko und Mittelamerika heimisch, und da sie auf natürliche Weise nur von bestimmten Bienen und Kolibris bestäubt werden kann, existierte die „Schwarze Blume der Azteken" lange nur in diesen Regionen. Über den spanischen Eroberer Hernán Cortés kam im Zuge der Kolonialisierung und Ausbeutung der Azteken zu Beginn des 16. Jahrhunderts die Vanille nach Europa: Geboren war das exklusive Gewürz der Könige und Reichen. Erst als 1841 der zwölfjährige Sklavenjunge Edmond Albus auf La Réunion – früher Île Bourbon genannt – entdeckte, wie die Blüte der inzwischen dorthin eingeschmuggelten Orchidee manuell bestäubt werden konnte, begann die Kultivierung der Bourbon-Vanille im großen Stil.

Orchis morio and associates

SALEP AND MARAŞ ICE CREAM

TURKEY IS HOME TO HEAVENLY DRINKS AND ICE CREAM MADE FROM ORCHID TUBERS

The root tubers of the orchid have been traded as an aphrodisiac since the Middle Ages. This may be why *salep*, a warming drink that was already popular in the Ottoman Empire during the winter months, remains popular today. The powder is made by boiling and drying the root tubers of various marsh orchids, mainly *Orchis morio*, and is mixed with milk, sugar, and spices to create a delicious hot drink that was once believed to have immune-boosting properties.

Salep remains a popular summer delight, too. The powder is a crucial ingredient in the famous Turkish ice cream known as *Maraş dondurması*, giving it a distinctive, slightly chewy texture. Hailing from the Kahramanmaraş region, this stretchy delicacy doesn't melt quite so readily in the heat while bustling street vendors deftly extract it from its containers in delightful, sticky strands onto long sticks and serve it to patrons.

Salep extraction poses a risk to the survival of these orchids in certain parts of Turkey since the process consumes the juicy root tubers that could otherwise sprout new flowers. In fact, it takes around 1,000 orchids to produce just one kilogram of *salep* powder, making them an endangered species in some areas.

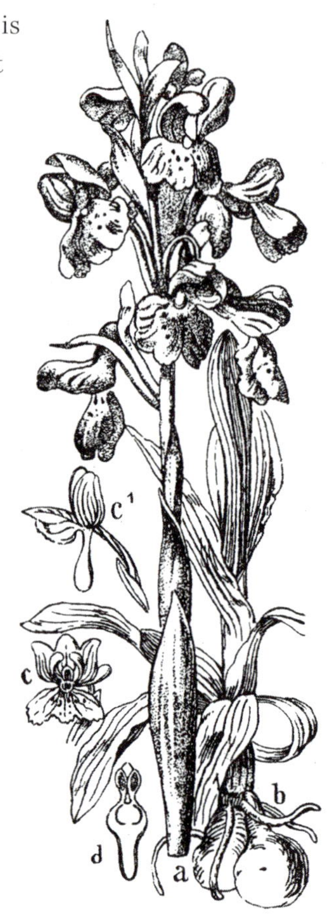

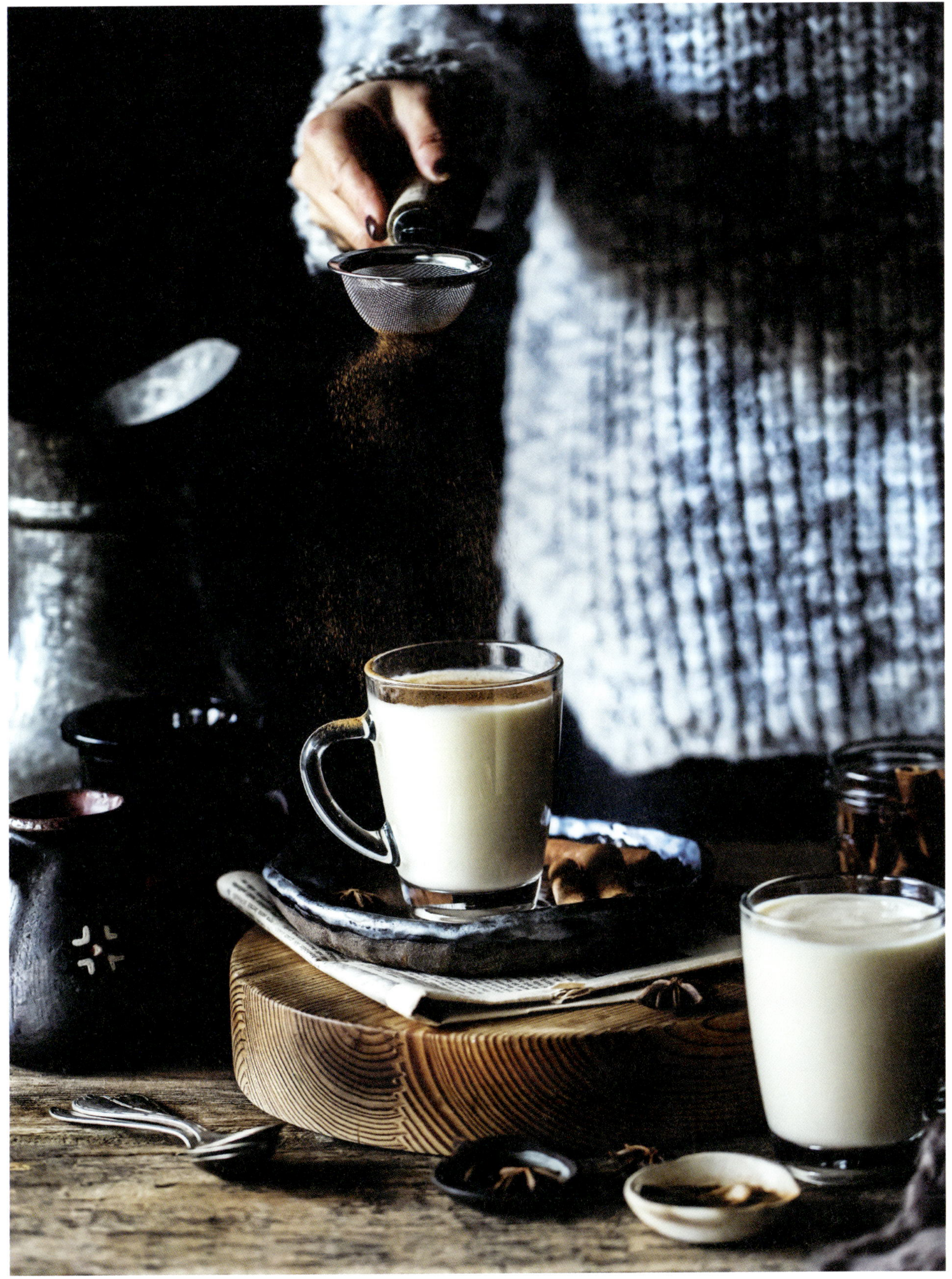

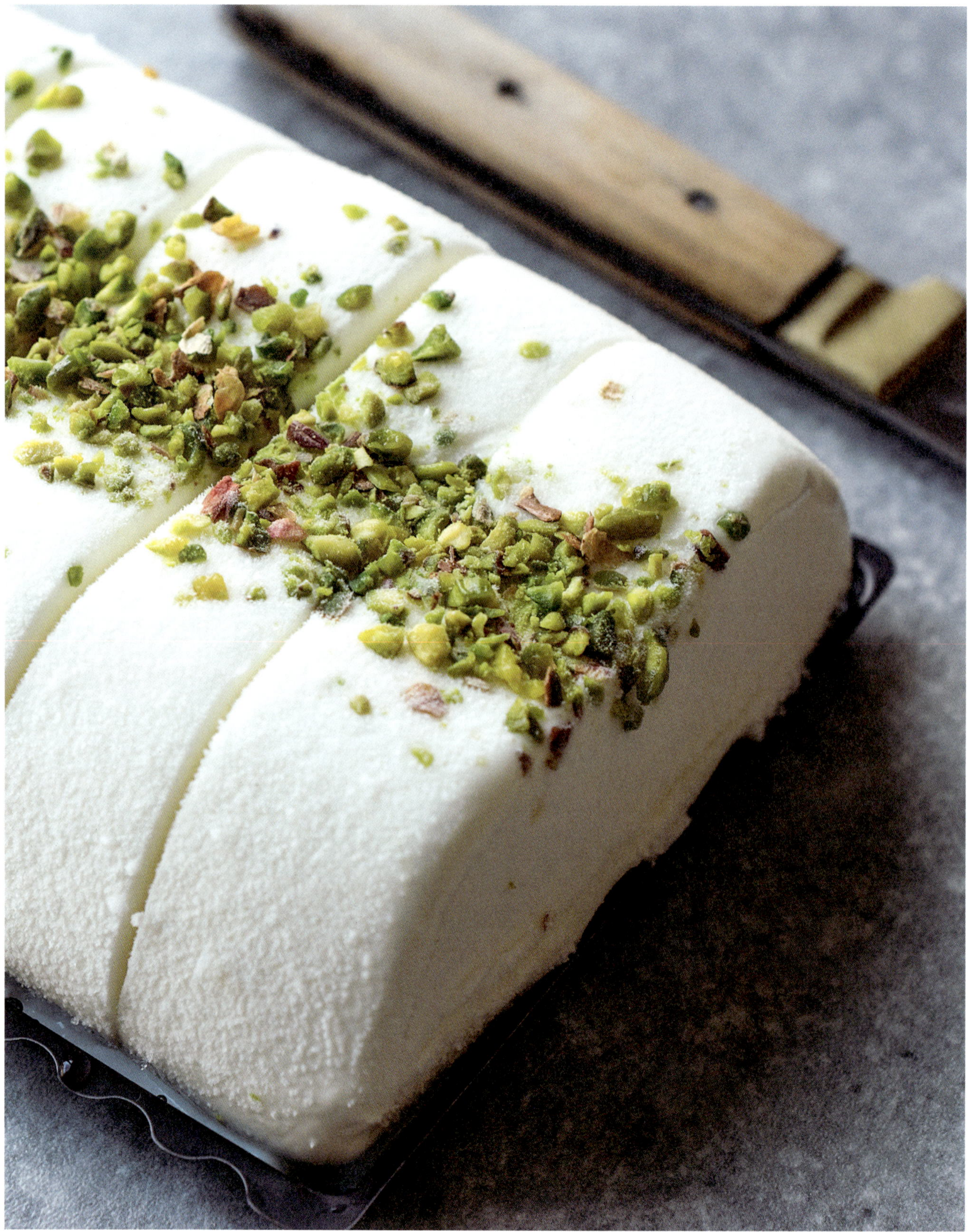

Orchis morio & Co.

SALEP UND MARAŞ DONDURMASI

KÖSTLICHE GETRÄNKE UND EIS AUS KNABENKRAUT-KNOLLEN FINDET MAN VORNEHMLICH IN DER TÜRKEI

Die Wurzelknollen des Knabenkrauts wurden schon im Mittelalter als Aphrodisiakum gehandelt. So erklärt sich vielleicht auch die Popularität von *Salep*, einem wärmenden Getränk, das bereits im Osmanischen Reich gerne in den Wintermonaten genossen wurde und sich auch heute noch großer Beliebtheit erfreut. Ein Pulver aus den gekochten und getrockneten Wurzelknollen verschiedener Knabenkräuter, vornehmlich *Orchis morio*, wird mit Milch, Zucker und Gewürzen zu einem leckeren Heißgetränk aufgekocht, dem früher eine immunstärkende Wirkung nachgesagt wurde.

Auch im Sommer steht *Salep* hoch im Kurs: Das Pulver ist ein wichtiger Bestandteil der berühmten türkischen Eisspezialität *Maraş dondurması* und verleiht dieser die charakteristische, etwas zähe Konsistenz. So schmilzt die elastische Leckerei aus der Region um die Stadt Kahramanmaraş auch bei höheren Temperaturen nicht, wenn geschäftige Straßenverkäufer sie in köstlich-klebrigen Strängen aus ihren Eisbehältern ziehen und an langen Stangen an ihre Kunden überreichen.

Da zur *Salep*-Gewinnnung die saftigen Wurzelknollen verwendet werden, aus denen sich wieder eine neue Blüte entwickeln könnte, gelten diese Knabenkräuter in einigen Regionen der Türkei heute als gefährdet: Für nur ein Kilogramm *Salep*-Pulver werden immerhin rund 1000 Orchideen benötigt!

ORCHIDS THROUGH HISTORY

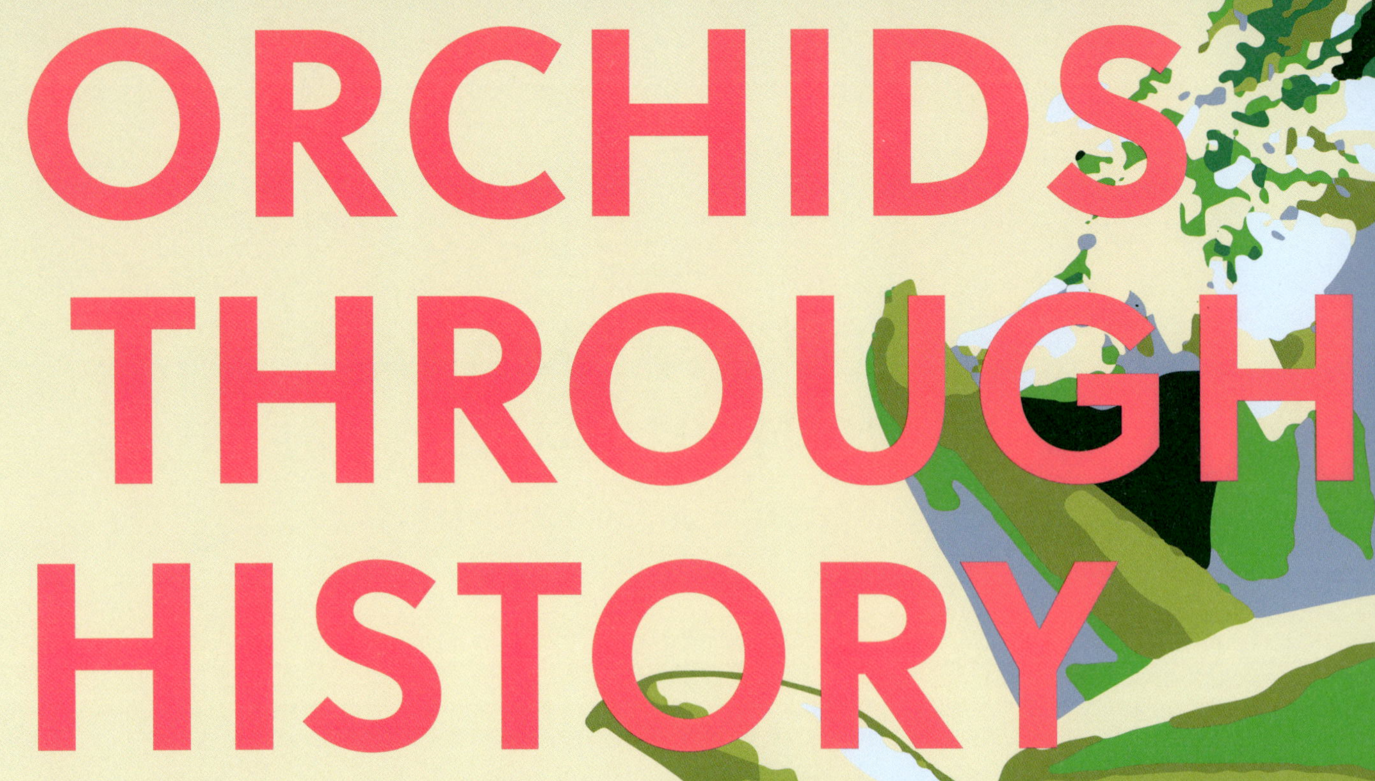

GESCHICHTE UND GESCHICHTEN

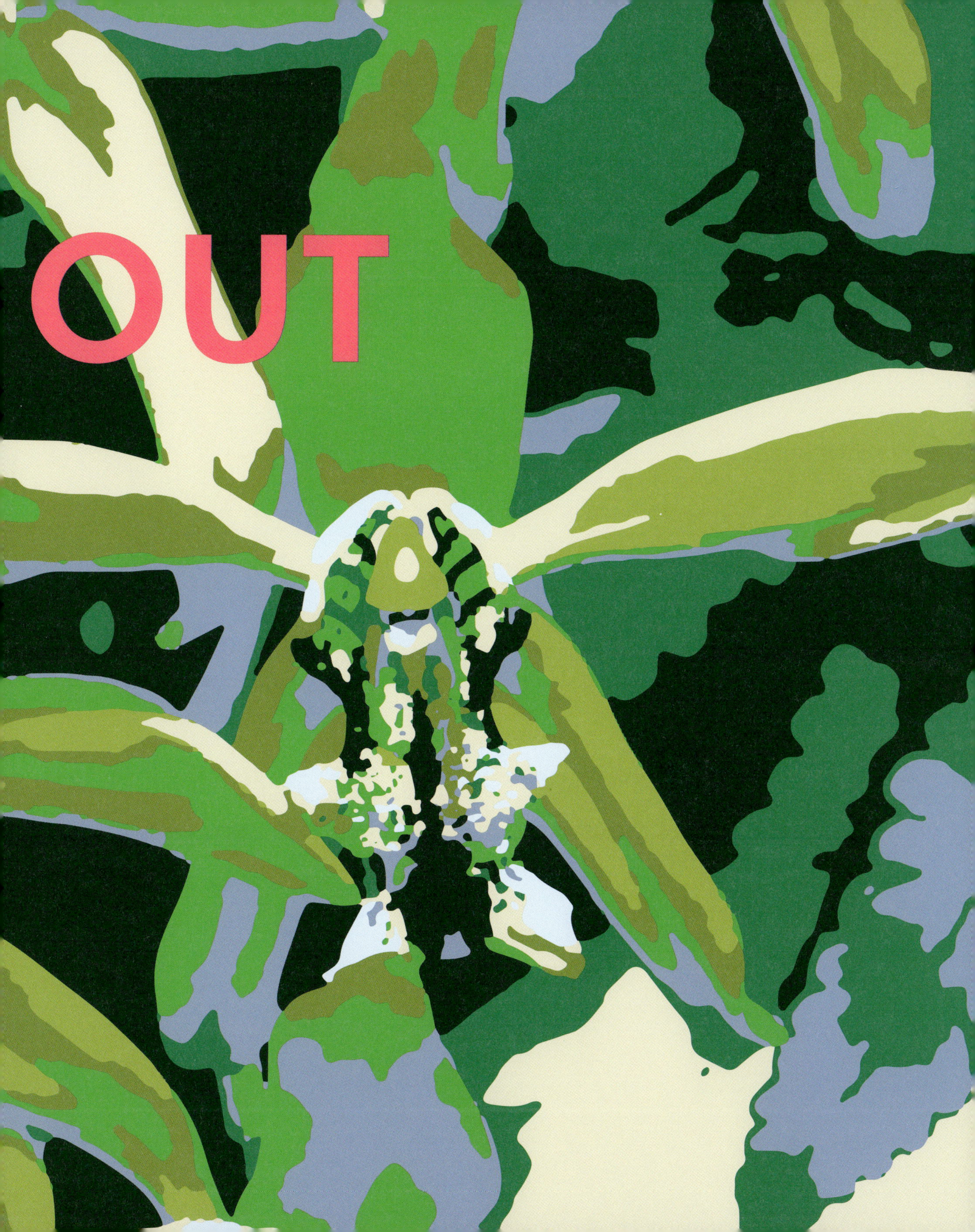
OUT

Orchis

HOW THE ORCHID GOT ITS NAME

AMAZINGLY, IT WAS NOT THEIR BEAUTIFUL FLOWERS BUT RATHER THEIR ROOTS

The Greek naturalist Theophrastus of Eresos was the first to mention orchids scientifically. In approximately 300 BCE, he used *orchis*, the Greek word for "testicles," to describe the genus, as the shape of their root tubers resembled that of male genitalia. This association also gave rise to the German common name for the genus, *Knabenkraut*, literally "knave-herb": it was a popular belief in Europe that consuming an *Orchis* root during pregnancy would increase the likelihood of giving birth to a boy.

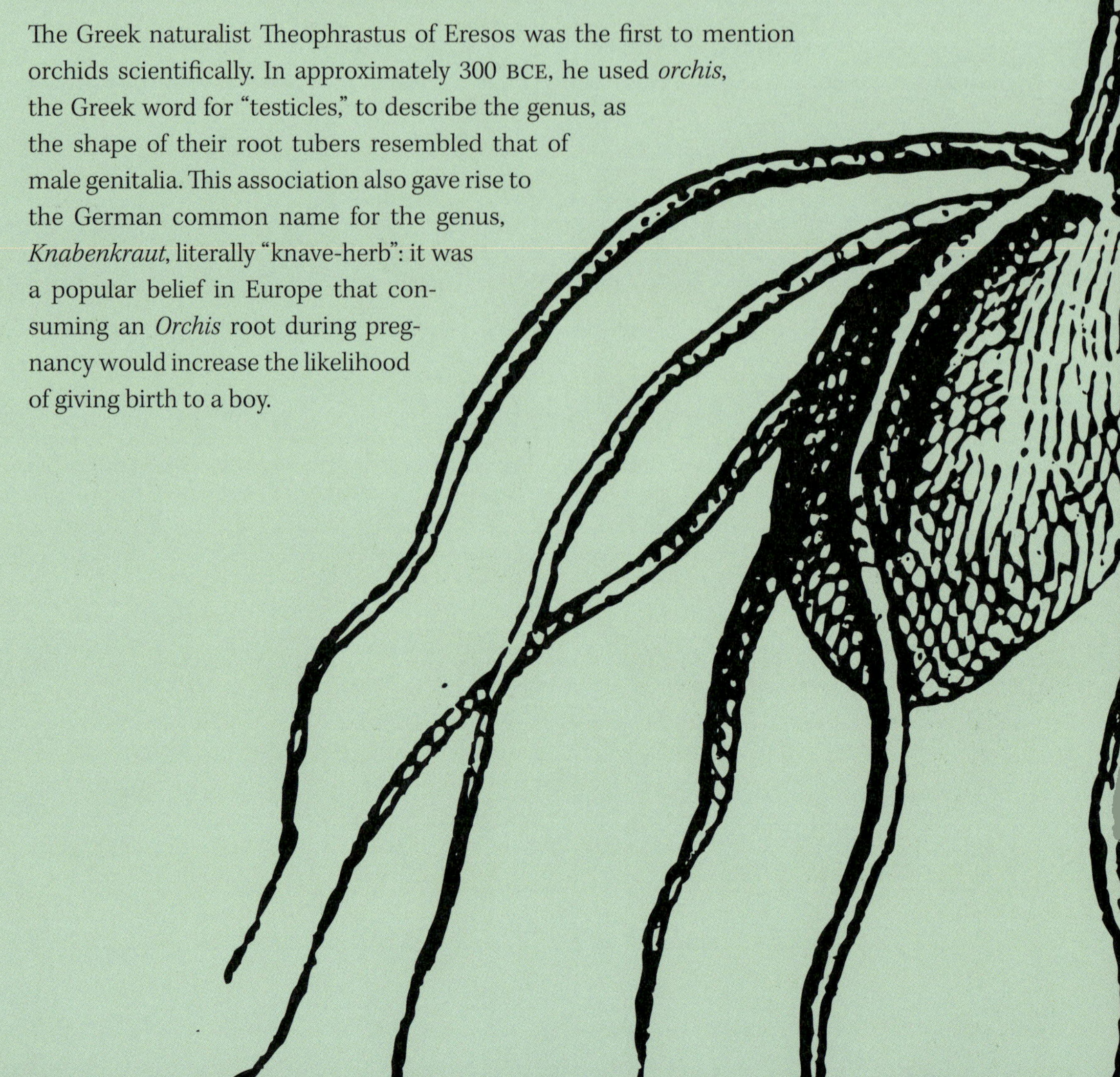

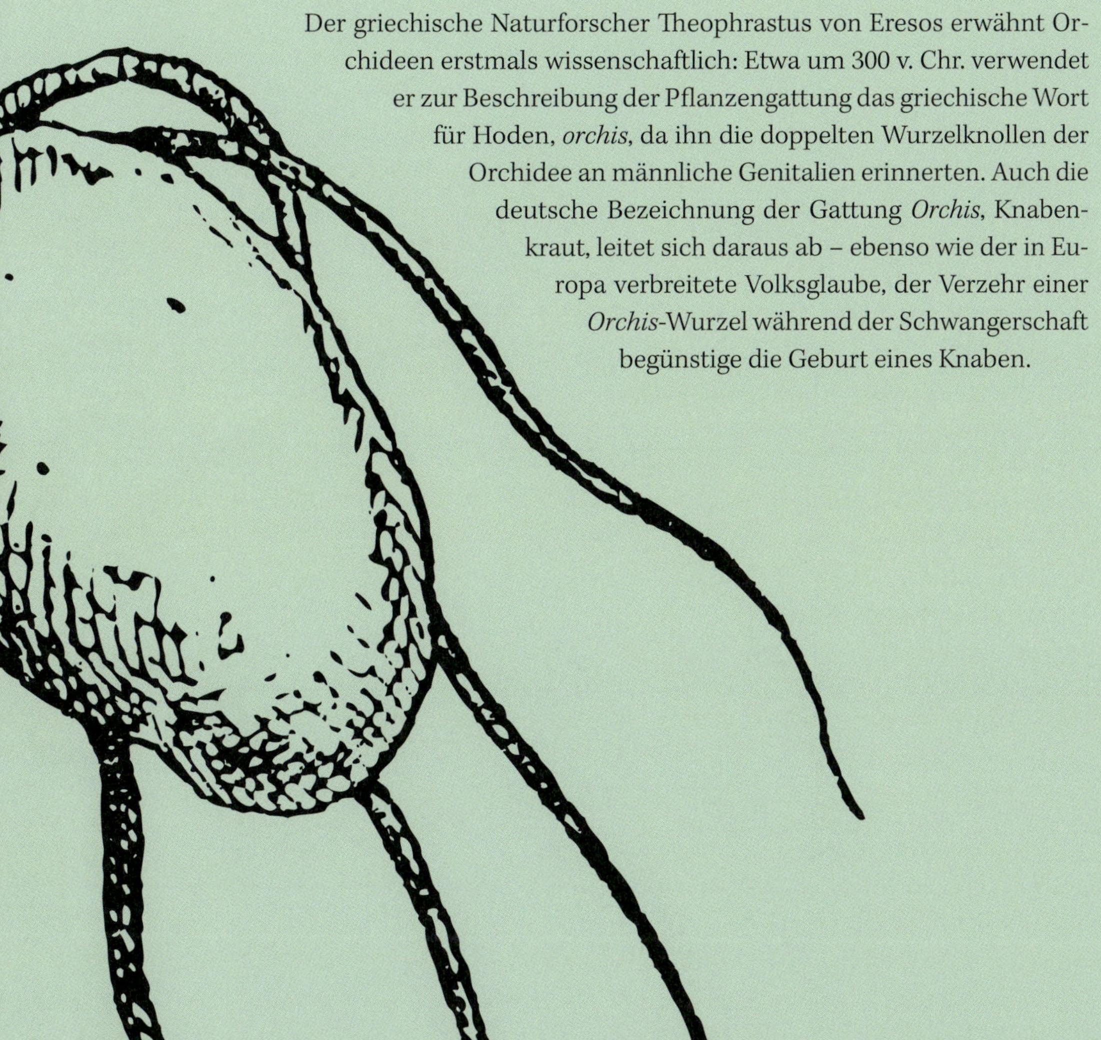

Orchis

DER NAME DER ORCHIDEE

ERSTAUNLICH, DASS NICHT DIE WUNDERSCHÖNE BLÜTE NAMENSGEBER DER ORCHIDEE WAR, SONDERN IHRE WURZEL

Der griechische Naturforscher Theophrastus von Eresos erwähnt Orchideen erstmals wissenschaftlich: Etwa um 300 v. Chr. verwendet er zur Beschreibung der Pflanzengattung das griechische Wort für Hoden, *orchis*, da ihn die doppelten Wurzelknollen der Orchidee an männliche Genitalien erinnerten. Auch die deutsche Bezeichnung der Gattung *Orchis*, Knabenkraut, leitet sich daraus ab – ebenso wie der in Europa verbreitete Volksglaube, der Verzehr einer *Orchis*-Wurzel während der Schwangerschaft begünstige die Geburt eines Knaben.

Eternal elegance

ORCHIDS, THE DINOSAURS OF THE FLORAL WORLD

A BEE ENCASED IN AMBER PROVIDES EVIDENCE OF THE ANCIENT ORIGINS OF ORCHIDS

Is it a direct line from the Big Bang to orchids? OK, not exactly. However, recent scientific research has revealed that orchids are far older than previously believed. Using the molecular clock method, biologists from Harvard University have determined that the first orchids appeared as long as 80 million years ago. This discovery was made possible by the identification of pollen from the *Meliorchis caribea* orchid species found on a bee preserved in amber.

Interestingly, there is written evidence of orchid cultivation dating back to 500 BCE, courtesy of China's most famous philosopher, Confucius. He was enamored by the beauty and fragrance of this queen of flowers, describing them as a symbol of grace, beauty, and sophistication. To this day, *lán*, the Chinese term for orchid, is used to denote refinement, purity, and elegance. For instance, one might refer to a cherished, trustworthy friend as one's "orchid friend".

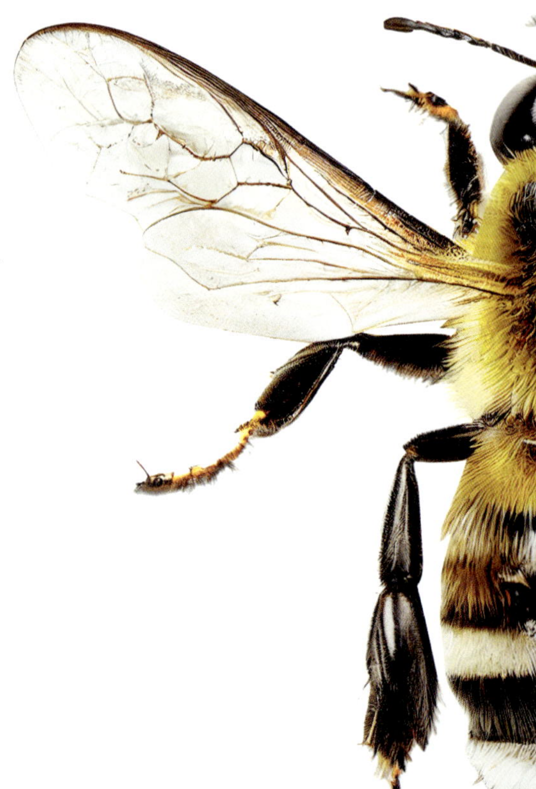

DIESE SEITE | On a bee preserved in amber, scientists found the pollen of orchids up to 80 million years old.

NÄCHSTE SEITE | Much more recently, orchids were mentioned in around 500 BCE by Confucius.

Zeitlose Schönheiten

DIE DINOSAURIER DER BLUMENWELT

EINE BIENE IN BERNSTEIN LIEFERT SPANNENDE ERKENNTNISSE ÜBER DAS WAHRE ALTER DER ORCHIDEEN

Vom Urknall direkt zur Orchidee? Nicht ganz, doch neueste wissenschaftliche Erkenntnisse zeigen, dass Orchideen viel älter sind, als wir bisher vermutet hatten. Biologen der Harvard-Universität datierten das Alter der ersten Orchideen nach der Methode der „molekularen Uhr" auf bis zu 80 Millionen Jahre: Möglich machte dies der Fund von Orchideenpollen der *Meliorchis caribea* an einer in Bernstein konservierten Biene. Schriftliche Beweise für die Kultur von Orchideen liefert uns übrigens Konfuzius etwa um 500 v. Chr.: Chinas wohl bekanntester Philosoph war sowohl von der Schönheit als auch vom Duft der „Königin der Blumen" fasziniert. Er beschreibt die Orchidee als Sinnbild von Anmut, Schönheit und Kultiviertheit. Auch heute wird *lán*, der chinesische Begriff für Orchidee, noch als Ausdruck des Feinen, Reinen und Eleganten verwendet: So könnte man einen ganz besonderen, aufrichtigen Freund beispielsweise auch als seinen „Orchideen-Freund" beschreiben.

DIESE SEITE | Auf einer in Bernstein eingeschlossenen Biene wurden Orchideenpollen gefunden – bis zu 80 Millionen Jahre alt.

NÄCHSTE SEITE | Vergleichsweise junger Beleg: Um 500 v. Chr. hat Konfuzius Orchideen erwähnt.

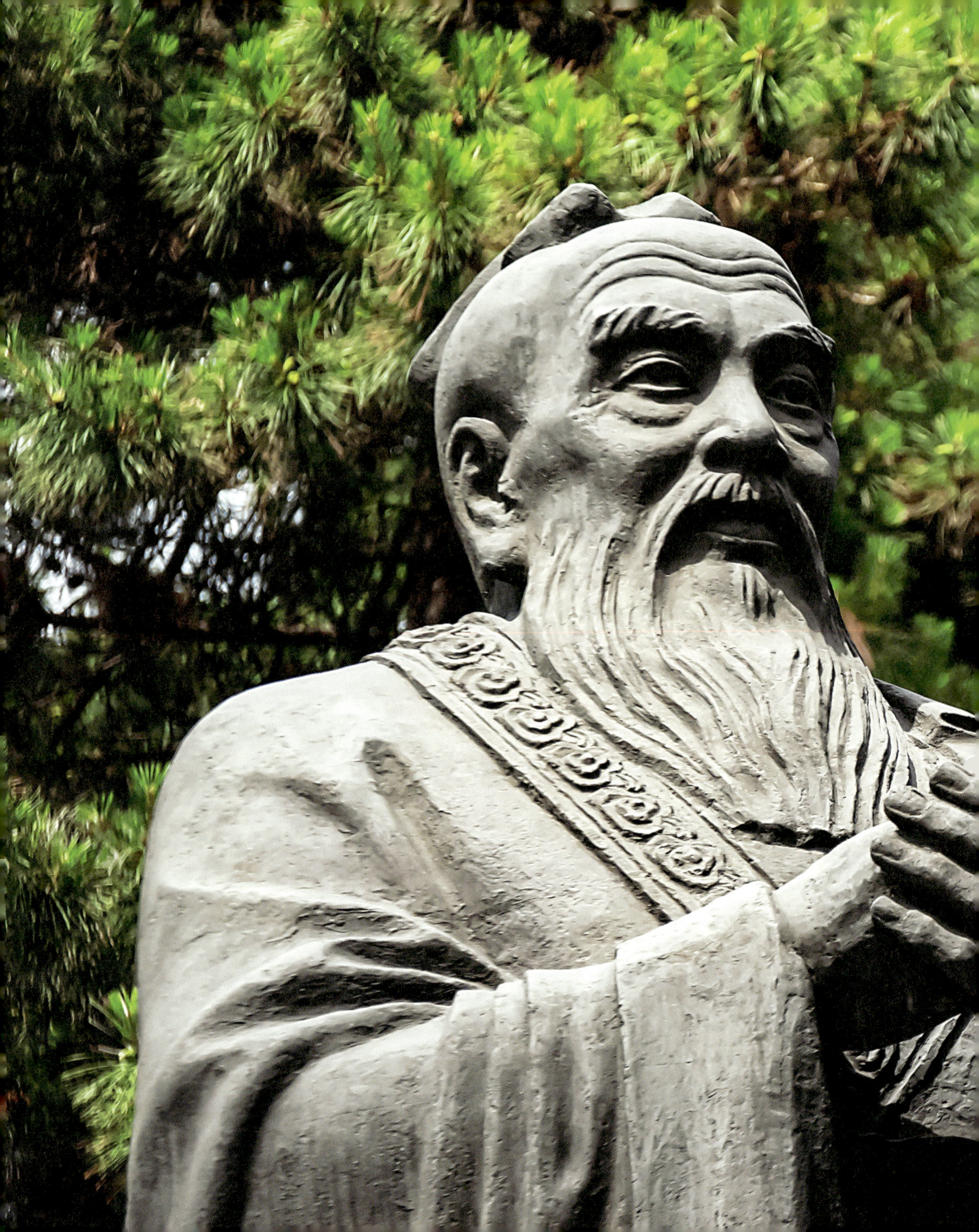

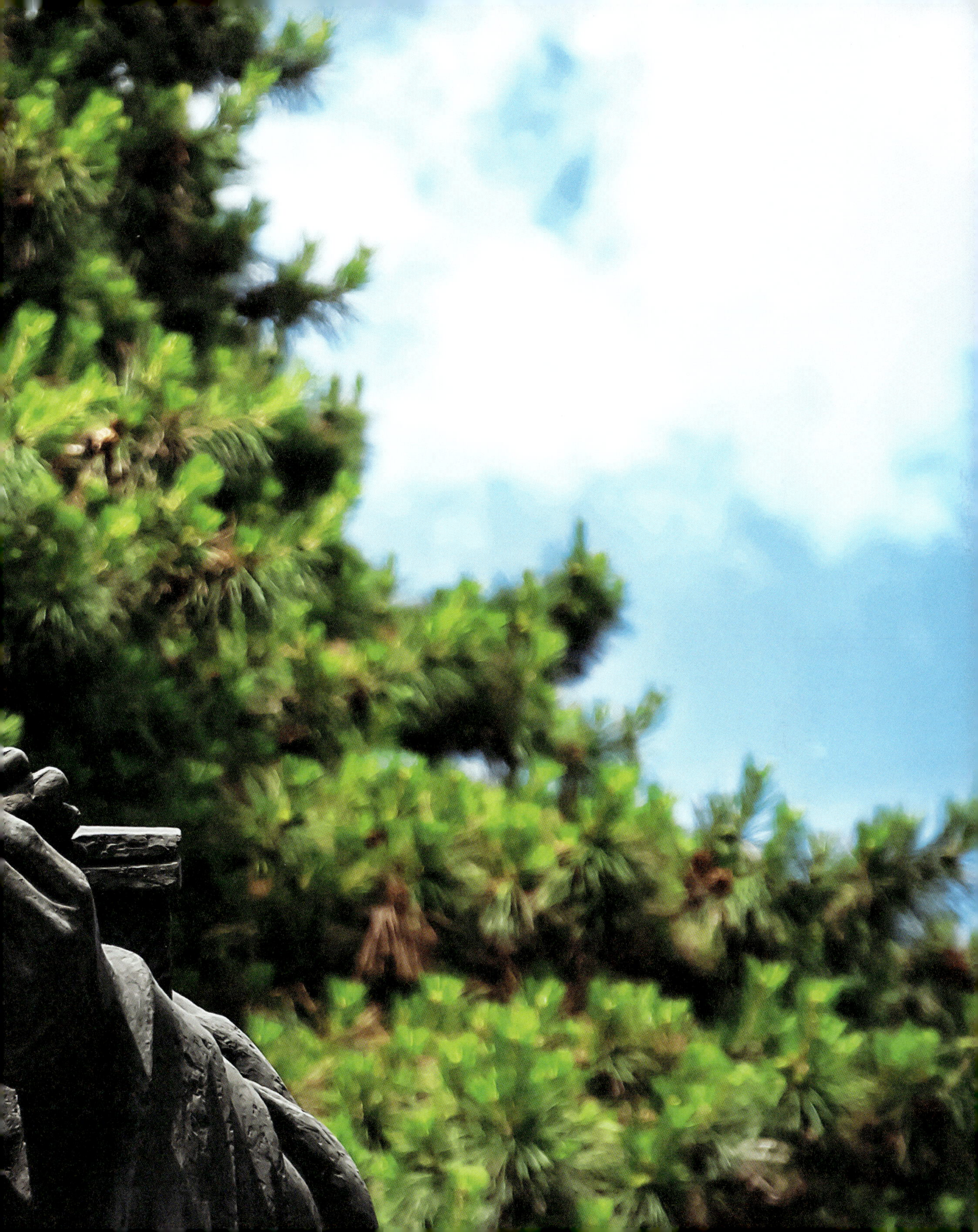

The world succumbs to orchid fever

ORCHIDELIRIUM

A VICTORIAN-ERA ADDICTION TO NEW, EXOTIC SPECIMENS BORE STRANGE FRUITS

Orchidelirium: It almost sounds like a disease, doesn't it? Well, in fact, the nineteenth-century British orchid collectors were so pathologically obsessed with these plants that the term "orchid fever" was coined to describe this peculiar phenomenon. The rarer and more exotic the plants that landed in the greenhouses of the Victorian elite, the more prestige and honor their wealthy owners could enjoy. These collectors would often spend ungodly sums on orchid hunters' expeditions, particularly to South and Central America, in the hope of discovering a new, unique species that could be named after them. It was the perfect hobby for affluent "gentlemen" who already had everything but were still seeking something more.

Orchidomania in England is said to have begun in 1818 when William John Swainson sent some tropical plants from Rio de Janeiro to a London merchant and orchid enthusiast named William Cattley. When Cattley opened the chest, he was surprised to find a fascinating purple flower of *Cattleya labiata* in full bloom. This incident led to the myth that the orchid had been used as packing material in the crate, but Swainson's botanical expertise makes this highly unlikely.

THIS PAGE | *Ophrys speculum*, likely drawn by William John Swainson. Facing page: *Cattleya* on a stamp from Democratic Kampuchea

NEXT PAGE | *Cattleya labiata*

Die Welt im Orchideenfieber

ORCHIDELIRIUM

IN DER VIKTORIANISCHEN ZEIT TRIEB DIE SUCHT NACH NEUEN EXOTISCHEN EXEMPLAREN SELTSAME BLÜTEN

Orchidelirium – klingt fast wie eine Krankheit: Tatsächlich trug es oft manische Züge, wie Orchideensammler im England des 19. Jahrhunderts den Objekten ihrer Begierde nachstellten, und so formte sich der Begriff des Orchideenfiebers für dieses ganz spezielle Zeitphänomen. Je seltener und reizvoller die exotischen Pflanzen waren, die in den Gewächshäusern der viktorianischen Haute-Volée landeten, umso mehr sonnten sich deren reiche Besitzer in Ruhm und Ehre. Sie waren gewillt, teils abstruse Summen in die Expeditionen der Orchideenjäger vornehmlich nach Süd- und Mittelamerika zu investieren. Immerhin lockte auch die Chance, mit einer auf den eigenen Namen getauften Neuentdeckung in die Geschichte einzugehen – das perfekte Hobby also für den reichen Gentleman, der schon alles besitzt, aber noch nicht genug hat …

Begonnen haben soll das englische Orchidelirium, als William John Swainson im Jahr 1818 tropische Pflanzen aus Rio de Janeiro an den Kaufmann und Orchideenkenner William Cattley nach London schickte. Beim Öffnen der Transportkiste entdeckte dieser unerwartet eine faszinierende purpurfarbene Blütenpracht. Dass diese *Cattleya labiata* aber nur als „Verpackungsmaterial" in die Kiste gestopft worden sein soll, gilt aufgrund der botanischen Expertise Swainsons heute aber als Mythos.

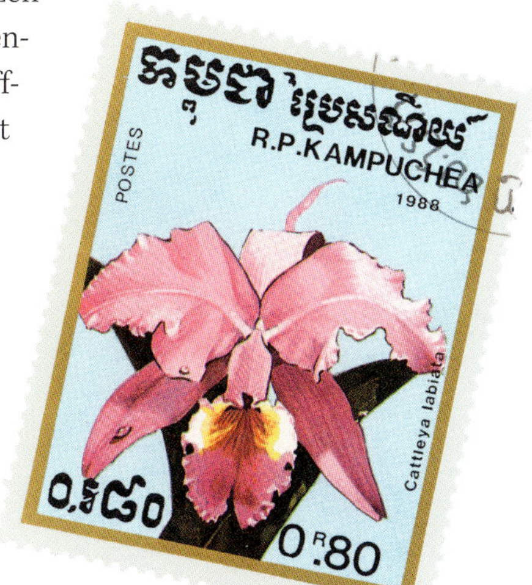

DIESE SEITE | *Ophrys speculum*, vermutlich von William John Swainson gemalt (links). Die *Cattleya* auf einer Briefmarke des Demokratischen Kampuchea (rechts)

NÄCHSTE SEITE | *Cattleya labiata*

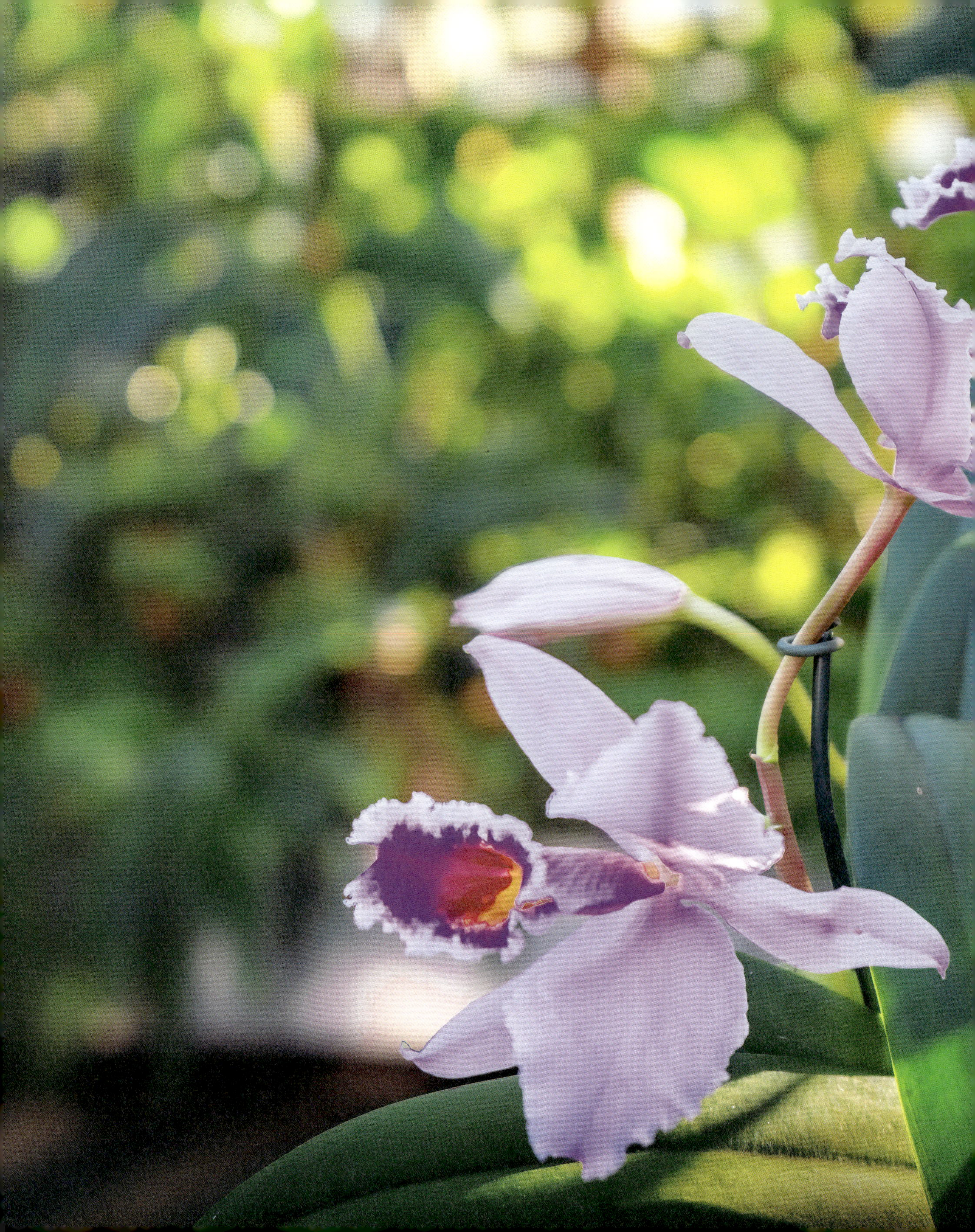

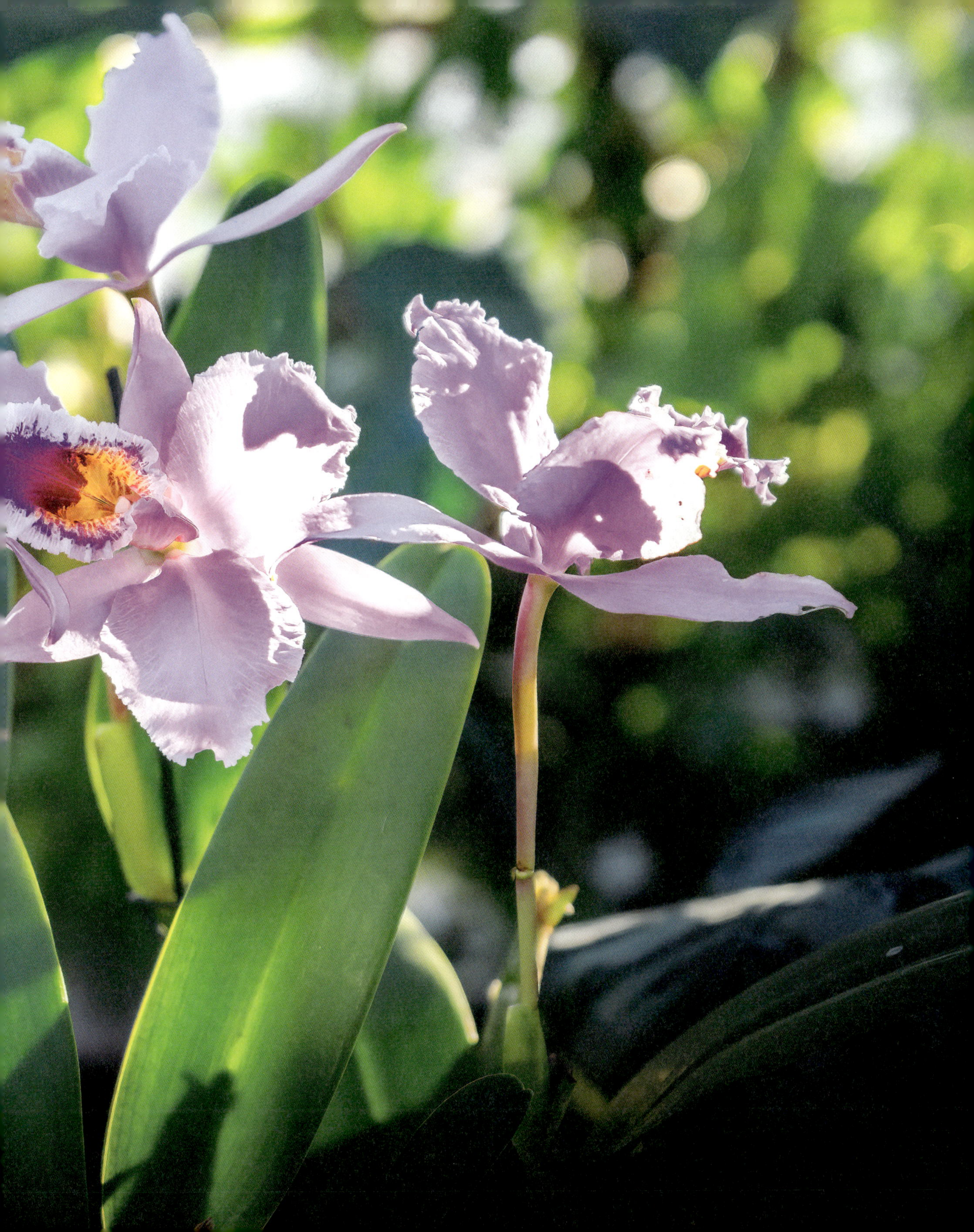

A touch of Indiana Jones

DARING ORCHID HUNTERS

NINETEENTH-CENTURY TALES OF ADVENTURE

During the Victorian era, orchid hunters embarked on a perilous life in pursuit of princely rewards. They braved tropical diseases, trekked through impassable rainforests and mountain ranges of South and Central America, and risked ambush by cutthroat competitors, who weren't beyond resorting to manslaughter or even murder. For instance, orchid hunter and painter Albert Millican wrote in his expedition reports from 1891 that he armed himself with various "knives, sabers, revolvers and rifles" to be prepared for an attack. These aren't exactly the tools one would expect from peaceful plant collectors.

Despite how childish it may sound, William Arnold, an orchid hunter, was allegedly ordered by his client, the Anglo-German orchid magnate Henry Frederick Sander, to urinate on his rivals' discoveries. If Arnold couldn't get ahold of the specimens himself, then nobody should.

NEXT PAGE | Left: A *Cattleya (hybrida) hardyana* painted by Henry George Moon for the *Reichenbachia* (see page 190). Right: An orchid in Colombia

Ein Hauch von Indiana Jones

WAGHALSIGE ORCHIDEENJÄGER

BERICHTE VON ABENTEURERN DES 19. JAHRHUNDERTS

Die viktorianischen Orchideenjäger *(siehe vorherige Seite)* ließen sich, mit der Aussicht auf eine fürstliche Entlohnung, auf ein gefahrvolles Leben ein: Neben tropischen Krankheiten und riskanten Routen durch die unwegsamen Regenwälder und Bergregionen Süd- und Mittelamerikas lauerten ihnen auch heimtückische Konkurrenten auf, die vor Mord und Totschlag nicht zurückschreckten. So schrieb Albert Millican, seines Zeichens Orchideenjäger und -maler, 1891 in seinen Expeditionsberichten, dass er sich mit einer Vielzahl von „Messern, Säbeln, Revolvern und Gewehren" bewaffnete, um gegen Angreifer gewappnet zu sein – nicht gerade die Ausrüstung, die man von friedliebenden Pflanzensammlern erwarten würde.

Da mutet es schon fast kindisch an, wenn berichtet wird, dass der Orchideenjäger William Arnold auf Geheiß seines Auftraggebers, des anglo-deutschen Orchideenmagnaten Henry Frederick Sander, auf die Fundstücke der Rivalen urinieren sollte: Wenn er dieser Orchideen schon nicht habhaft werden konnte, sollte auch kein anderer die Schätze nach Hause bringen!

NÄCHSTE SEITE | Links: Eine *Cattleya (hybrida) hardyana*, gezeichnet von Henry George Moon für die *Reichenbachia* (siehe Seite 191). Rechts eine Orchidee in Kolumbien

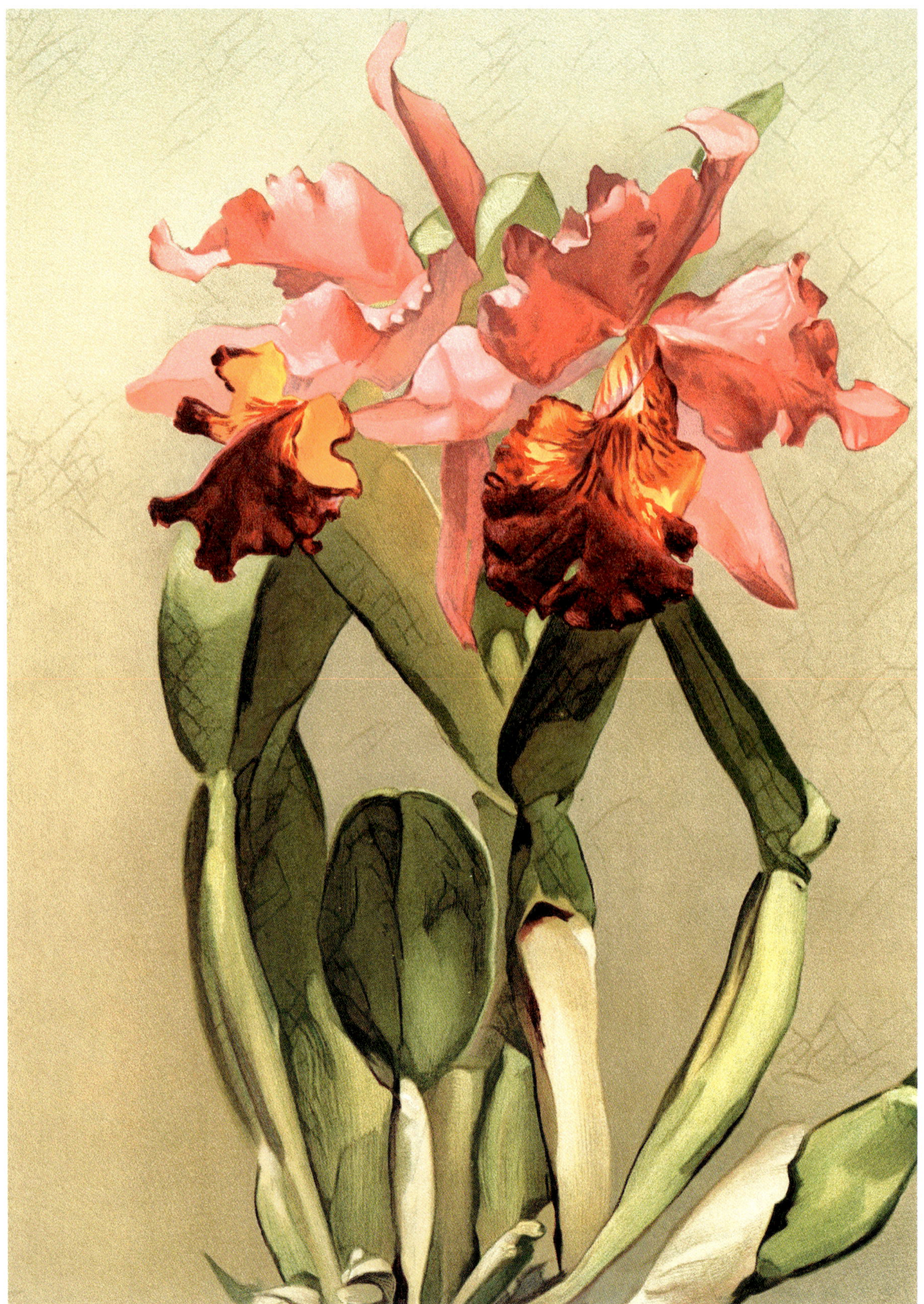

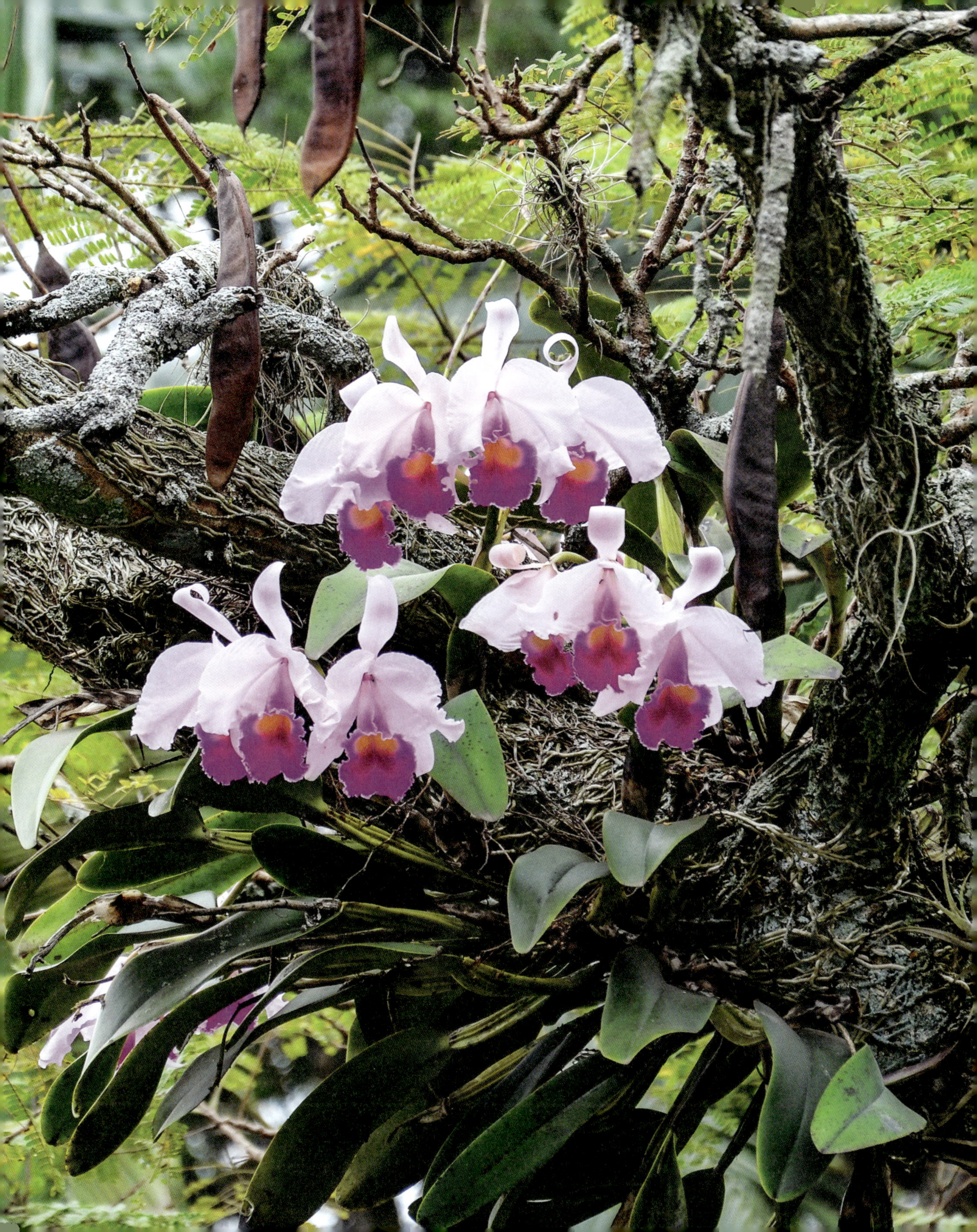

Renowned expert and publicist

JAMES BATEMAN

AN ORCHID COLLECTOR WITH BIG IDEAS

Wealthy, upper-class men during the Victorian era, like James Bateman (1811–1897), were the driving force behind orchid-omania. As a landowner and botanist, Bateman sponsored dangerous orchid expeditions to Central America and created what may have been England's most impressive orchid collection on his country estate. He conducted meticulous studies of the conditions in which the plants thrived, and the knowledge he gained helped him grow *Odontoglossum* in England, not by relying on tropical conditions in orchid houses, but by simulating the cool climate of the South American areas from which these plants originate.

Remarkably, Bateman was only 26 years old when he began his richly illustrated work on the orchids of Mexico and Guatemala. To this day, his standard work is also one of the largest orchid publications ever, measuring a stately 30 x 22 inches, or about 76 x 56 centimeters. Perhaps it was fortunate that only 125 copies of *The Orchidaceae of Mexico and Guatemala* were printed; even a contemporary of Bateman's, a cartoonist named George Cruikshank, lampooned this superlative volume as every librarian's nightmare.

THIS PAGE| A *Galeandra baueri* as represented by James Bateman

NEXT PAGE | An orchid of the genus *Odontoglossum* found in central and western South America

Anerkannter Experte und Publizist

JAMES BATEMAN

EIN ORCHIDEENSAMMLER, DER GROSSES IM SINN HATTE

Treibende Kräfte des Orchideliriums waren Mitglieder der reichen viktorianischen Oberschicht wie James Bateman (1811–1897). Der vermögende Landbesitzer und Botaniker war Sponsor waghalsiger Orchideenexpeditionen nach Mittelamerika und baute auf seinem Landsitz die vielleicht beeindruckendste Orchideensammlung Englands auf. Da er auch die Lebensbedingungen der Pflanzen aufmerksam studierte, gelang es ihm, *Odontoglossum* in England zu kultivieren, indem er nicht auf tropisch warme Orchideenhäuser setzte, sondern das kühle Klima der südamerikanischen Herkunftsgebiete imitierte.

Bemerkenswert: Schon im Alter von 26 Jahren begann James Bateman mit der Arbeit an seinem reich illustrierten Standardwerk über die Orchideen von Mexico und Guatemala, das bis heute zu den größten Orchideenpublikationen zählt – und zwar im ganz wörtlichen Sinn: Der Wälzer maß stattliche 30 auf 22 Zoll, also rund 76 mal 56 Zentimeter. Vielleicht ein Glück, dass es nur 125 Exemplare von *The Orchidaceae of Mexico and Guatemala* gab – denn der Band der Superlative wurde schon vom zeitgenössischen Karikaturisten George Cruikshank als „Albtraum jedes Bibliothekars" persifliert.

DIESE SEITE | Eine *Galeandra baueri* in der Darstellung von James Bateman

NÄCHSTE SEITE | Eine Orchidee der Gattung *Odontoglossum*, die im zentralen and westlichen Südamerika zu Hause ist

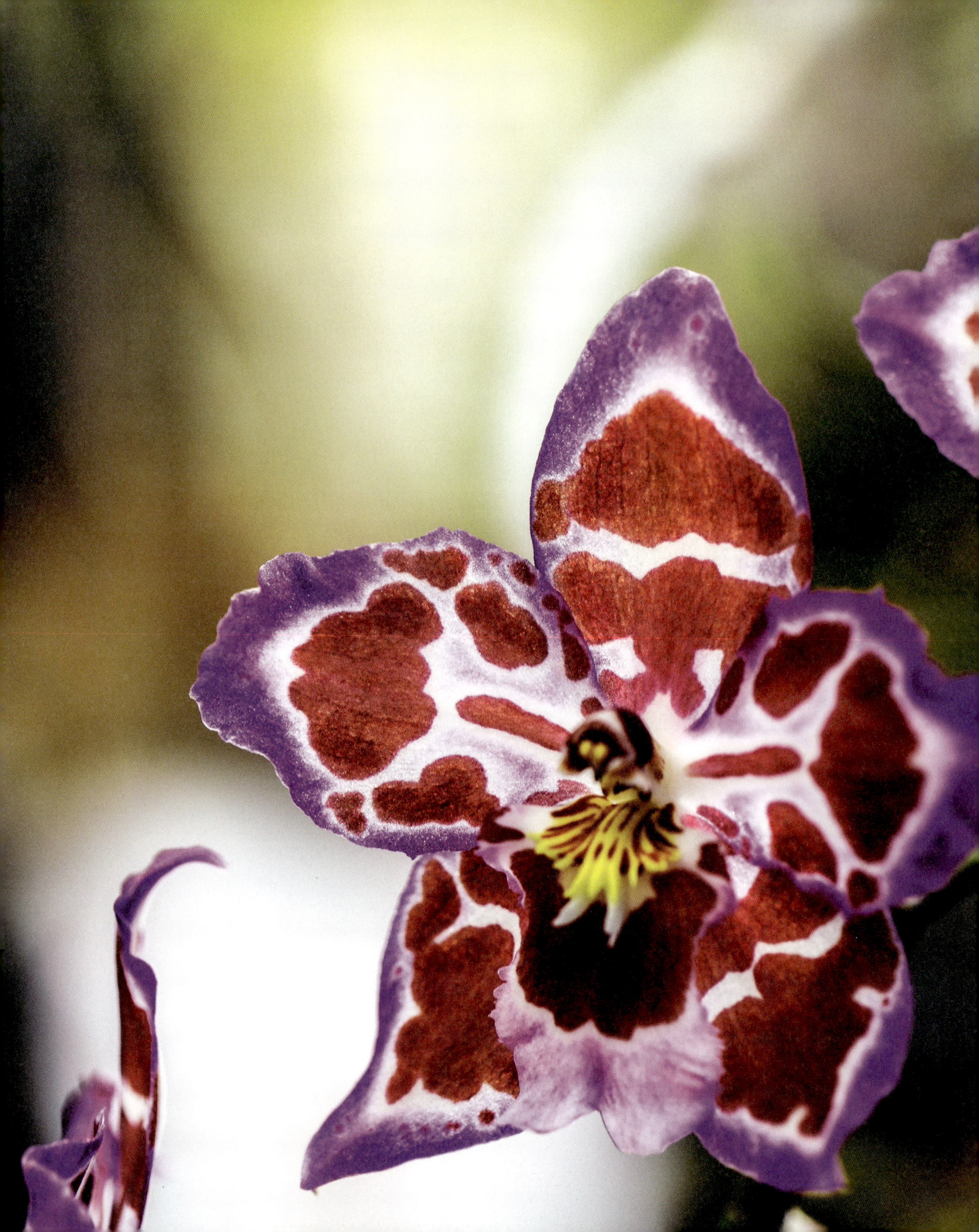

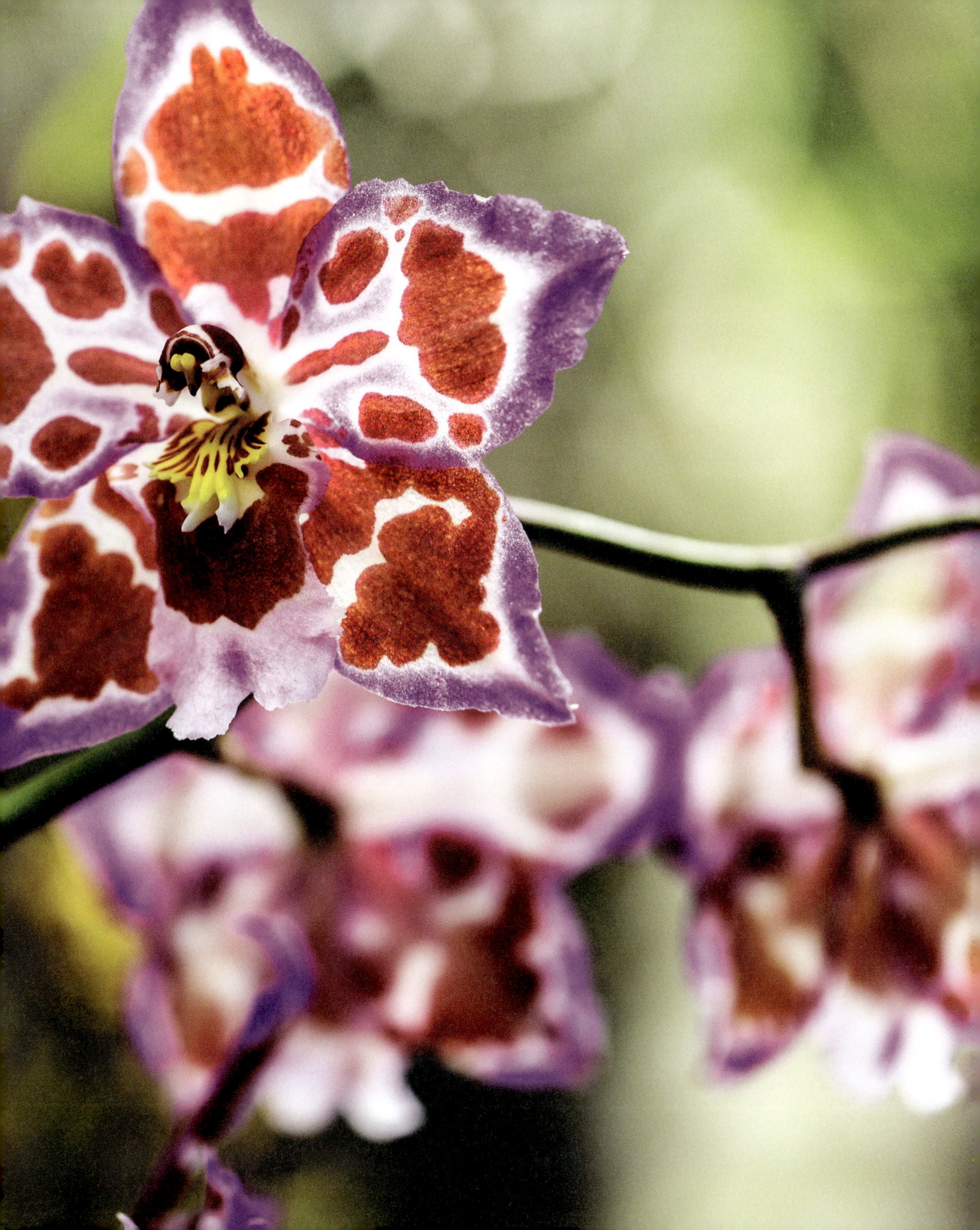

The founder of Orchidology

JOHN LINDLEY

LINDLEY AUTHORED ONE OF THE NINETEENTH CENTURY'S MOST BEAUTIFUL ILLUSTRATED VOLUMES ON ORCHIDS

John Lindley (1799–1865), a British botanist, is credited with pioneering the modern study of orchids. He was the first to successfully classify the orchid species known at that time and named approximately 150 new species during his lifetime. His work was actively supported by the wealthy orchid collector William Cattley, whom he honored by naming after him the genus *Cattleya* as a sign of gratitude to his loyal friend and patron.

The son of a gardener, Lindley likely grew up with a deep love and respect for plants. He became an internationally recognized expert on orchids while working for the Royal Horticultural Society. In 1838, he published a tome called *Sertum Orchidaceum*, which featured highly artistic prints based on beautiful watercolors painted by Sarah-Ann Drake *(see page 196)*. The illustrations are extraordinarily rich and detailed. Notably, Lindley once played a crucial role in saving the royal greenhouses at Kew from imminent destruction. His botanical legacy lives on at Kew Gardens today.

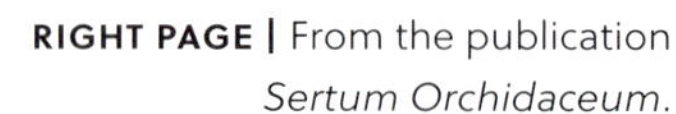

RIGHT PAGE | From the publication *Sertum Orchidaceum.*

„Vater der Orchideenkunde"

JOHN LINDLEY

VOM IHM STAMMT EINES DER SCHÖNSTEN ILLUSTRIERTEN ORCHIDEENWERKE DES 19. JAHRHUNDERTS

Der britische Botaniker John Lindley (1799–1865) gilt als Pionier der modernen Orchideenkunde: Ihm gelang es erstmals, die damals bekannten Orchideenarten zu klassifizieren und zu seinen Lebzeiten rund 150 neue Arten zu benennen. In seiner Arbeit wurde er tatkräftig durch den wohlhabenden Orchideensammler William Cattley unterstützt: Nicht zuletzt durch die Benennung der Gattung *Cattleya* sprach er seinem treuen Freund und Gönner dafür seinen Dank aus.

Als Sohn eines Gärtners war ihm die Pflanzenliebe wohl in die Wiege gelegt. Während seiner Tätigkeit für die Royal Horticultural Society wurde er zum international anerkannten Spezialisten für Orchideengewächse: 1838 veröffentlichte er den Orchideen-Folianten *Sertum Orchidaceum*, außergewöhnlich detailreich illustriert mit kunstvollen Drucken nach den wundervollen Aquarellen von Sarah-Ann Drake *(siehe Seite 197)*. Übrigens ist es auch John Lindleys Engagement zu verdanken, dass die königlichen Gewächshäuser in Kew vor der drohenden Zerstörung bewahrt wurden – und folgerichtig lebt sein botanisches Erbe auch heute noch dort weiter.

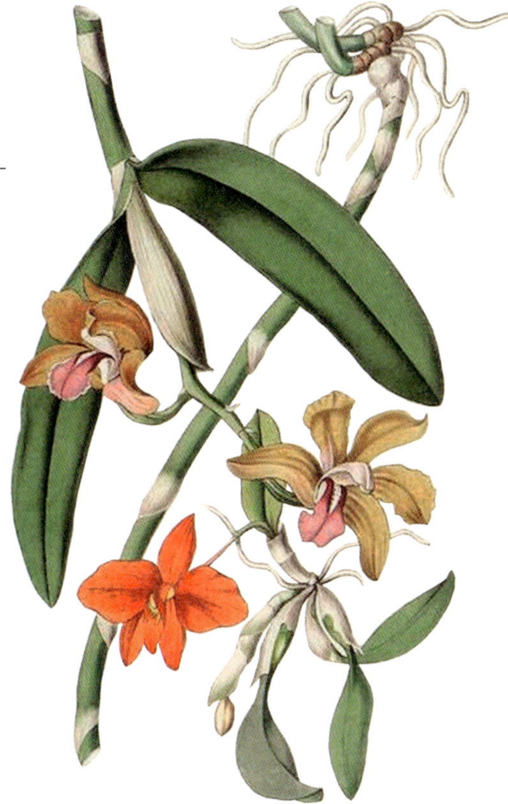

DIESE SEITE | Darstellung aus dem Band *Sertum Orchidaceum*.

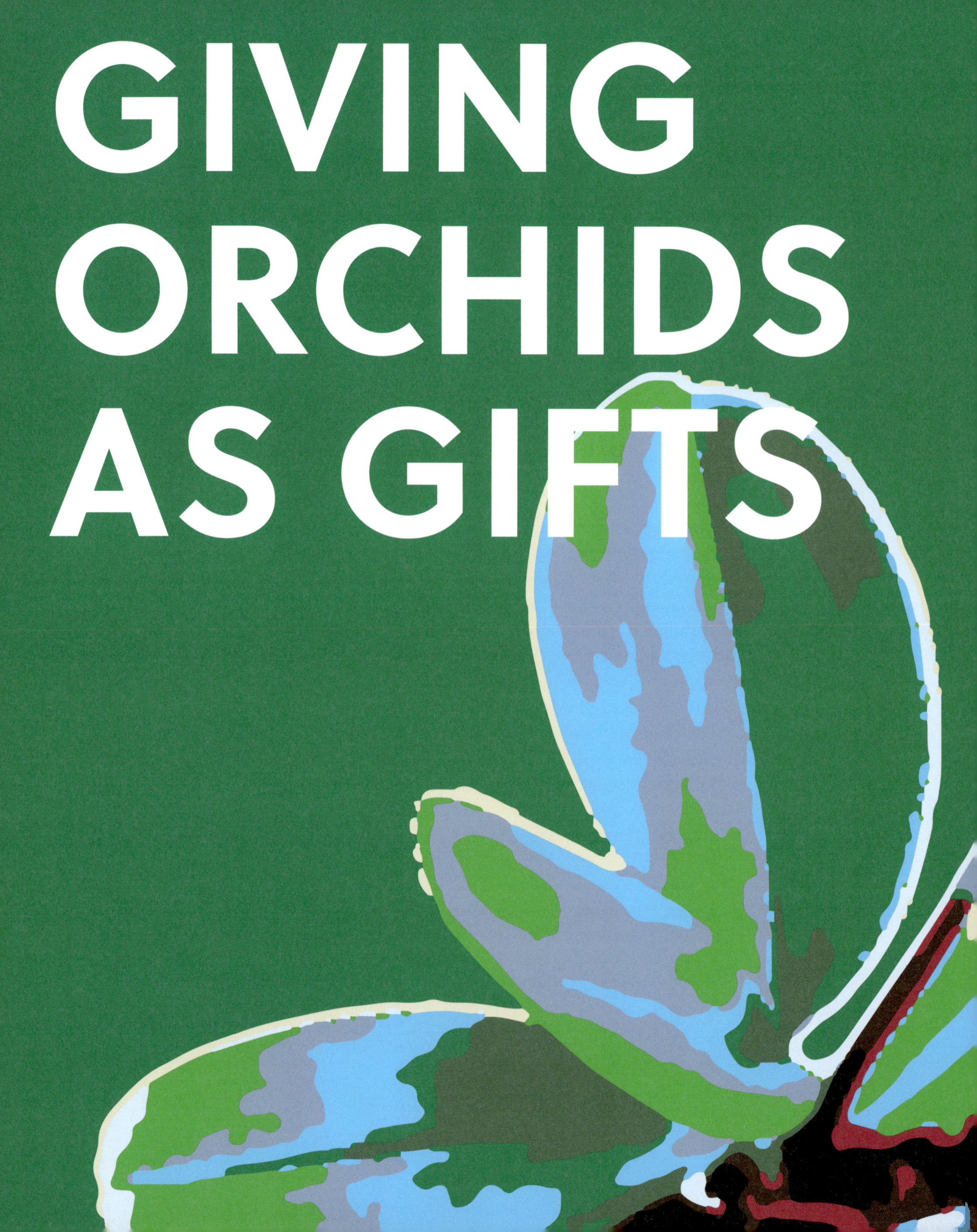
GIVING
ORCHIDS
AS GIFTS

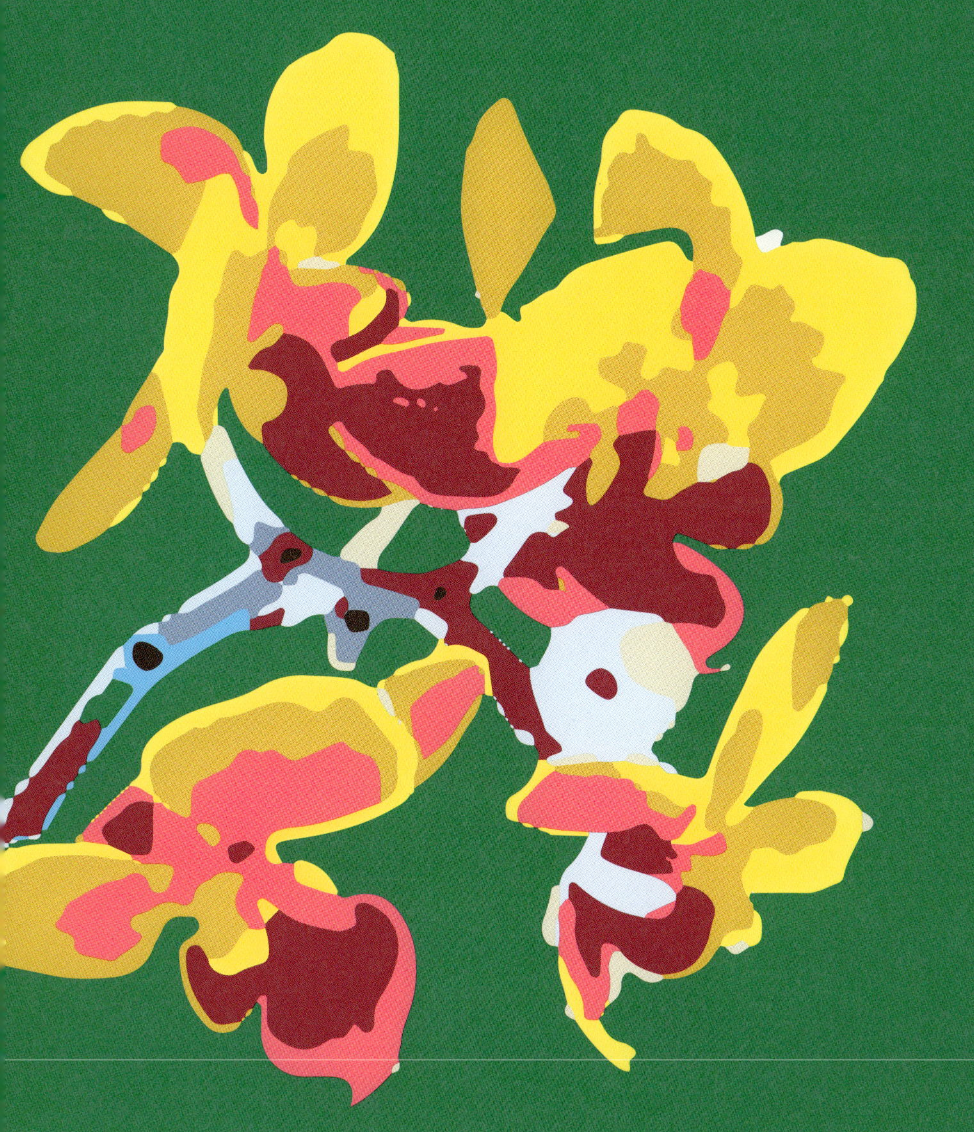

ORCHIDEEN
SCHENKEN

Color and meaning

AN EASY GUIDE: GIVING ORCHIDS AS GIFTS

WHEN IT COMES TO FLOWERS, COLOR SETS THE RIGHT TONE – EVEN IN THE WORLD OF ORCHIDS

Few flowers have been associated with as many symbolic meanings over time as the orchid. With their exotic shapes, colors, and alluring fragrances, orchids have become a symbol of beauty, passion, fertility, and masculinity. They're also associated with extravagance and nobility, which is why they're often seen as a symbol of success and wealth. If you give orchids as a gift, you can't go wrong – especially if you know their secret code of colors.

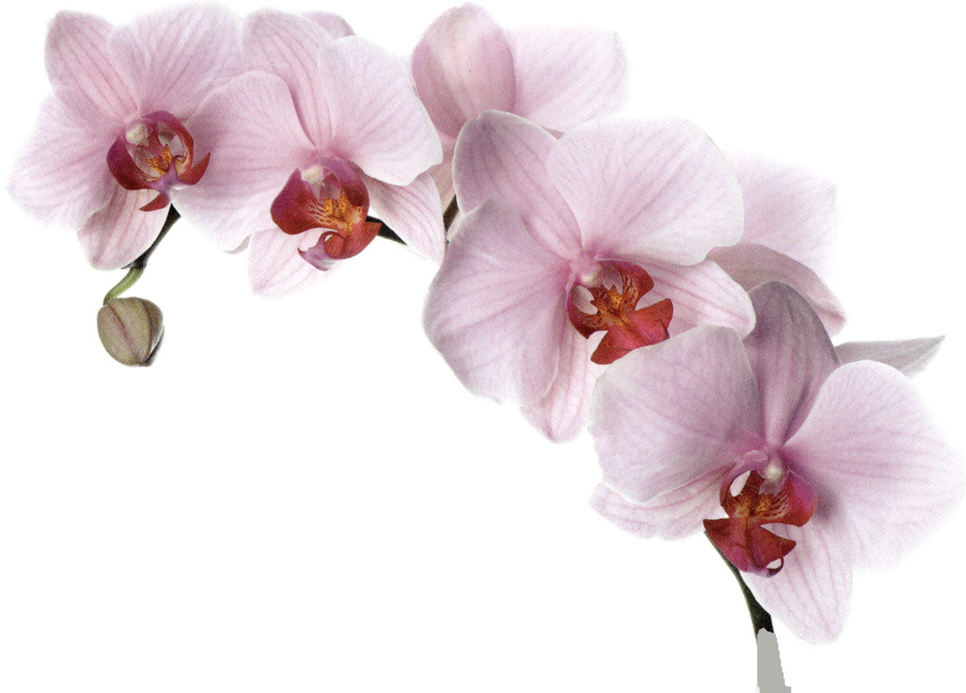

Farbe und Bedeutung

ORCHIDEEN SCHENKEN LEICHT GEMACHT

WENN BLUMEN SPRECHEN, GIBT DIE FARBE DEN TON AN – AUCH IN DER WELT DER ORCHIDEEN ...

Kaum eine Blume bekam im Laufe der Zeit so viele symbolhafte Bedeutungen zugeschrieben wie die Orchidee: Ihre exotischen Blütenformen, Farben und verführerischen Düfte ließen Orchideen zum Symbol für Schönheit, Leidenschaft, Fruchtbarkeit und Männlichkeit werden. Auch der Hauch des Extravaganten und Edlen begleitet sie, und so verwundert es nicht, dass die „Königin der Blumen" auch für Erfolg und Reichtum steht. Wer Orchideen schenkt, liegt fast immer richtig – vor allem, wenn das Geheimnis der Farbensprache gelüftet ist ...

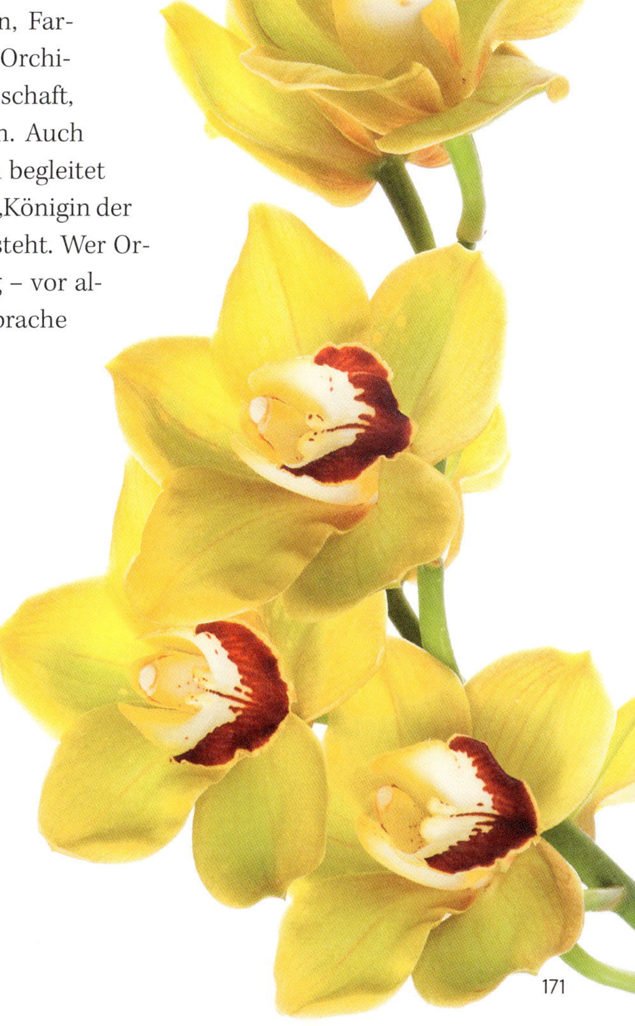

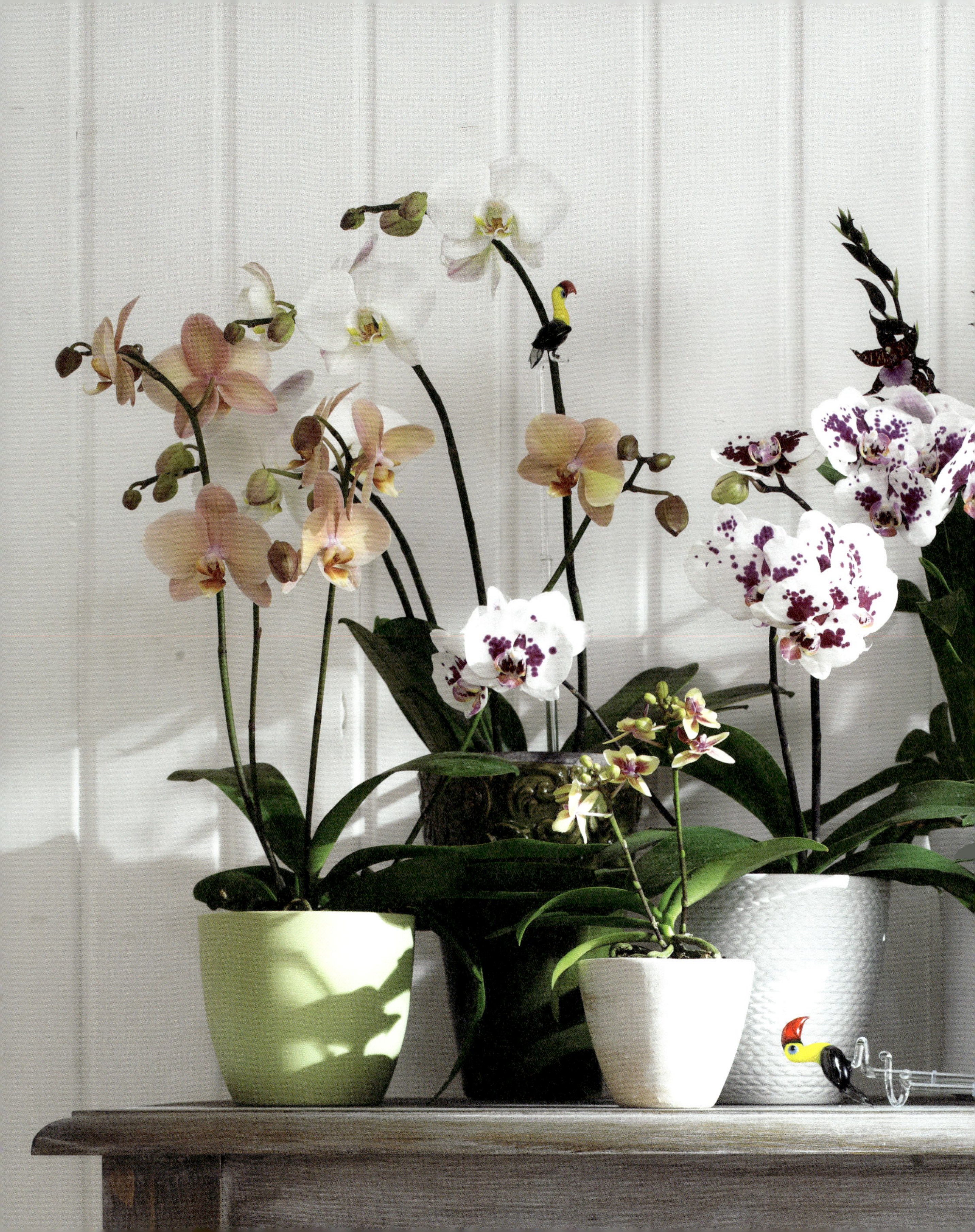

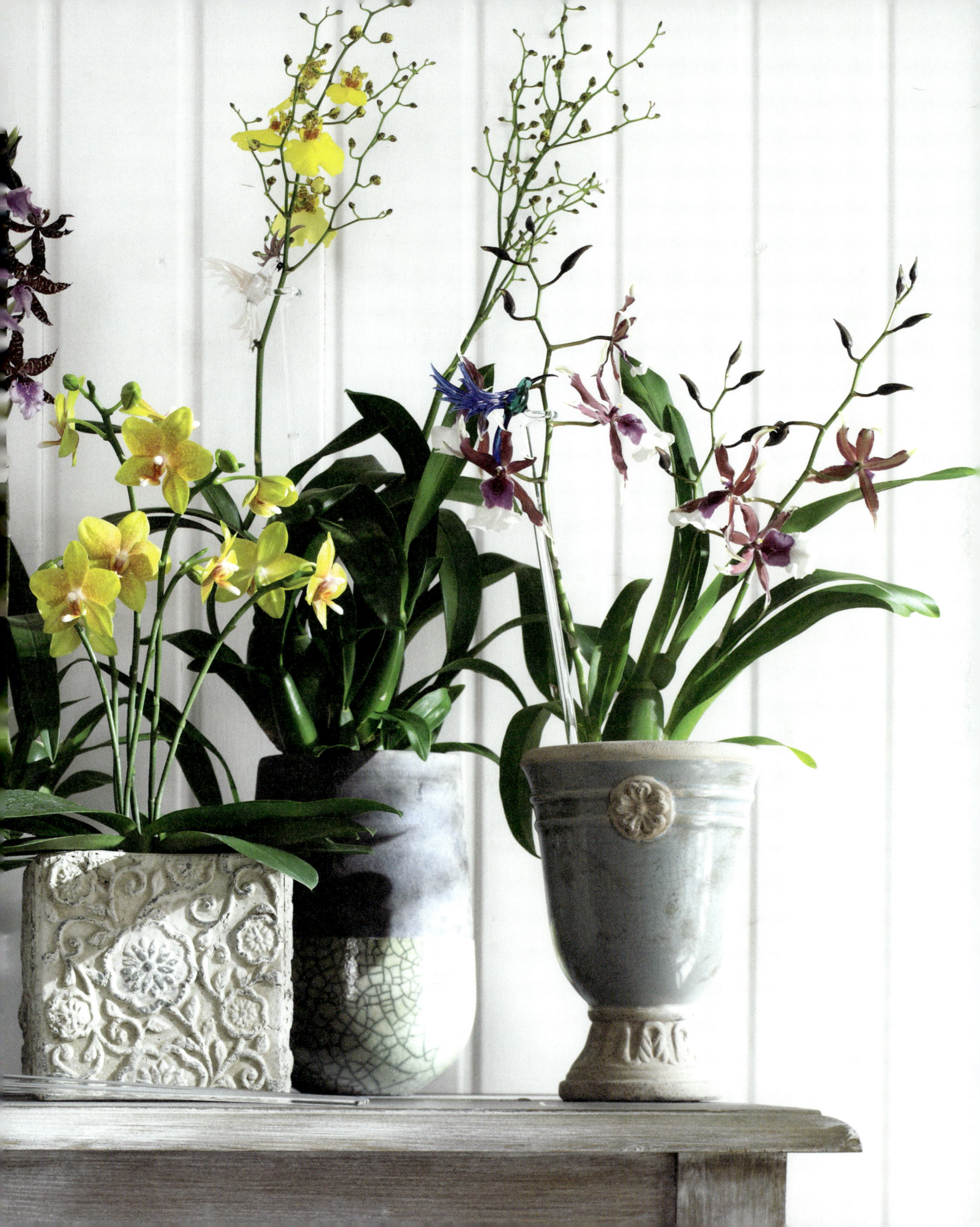

Cymbidium

THE COLOR OF PASSION

There's no doubt that when it comes to love and passion, red is the color that comes to mind. A red *Cymbidium* orchid is a perfect way to celebrate a wedding anniversary and show your partner that your passion burns just as brightly as it did on your first day of marriage. However, red orchids are also seen as a symbol of strength and courage. So, if your boss gives you a red orchid as a gift, don't necessarily take it as a romantic overture.

LEIDENSCHAFTLICHES ROT

Ganz klar: Wenn es eine Farbe gibt, die Liebe und Leidenschaft repräsentiert, ist es Rot. Wie wäre es mit einer roten *Cymbidium*, um den Hochzeitstag zu feiern und den Partner wissen zu lassen, dass die Leidenschaft noch so heiß glüht wie am ersten Tag? Rote Orchideen gelten aber auch als Zeichen von Stärke und Mut – wer vom Chef eine rote Orchidee geschenkt bekommt, sollte darin nicht unbedingt eine amouröse Avance sehen.

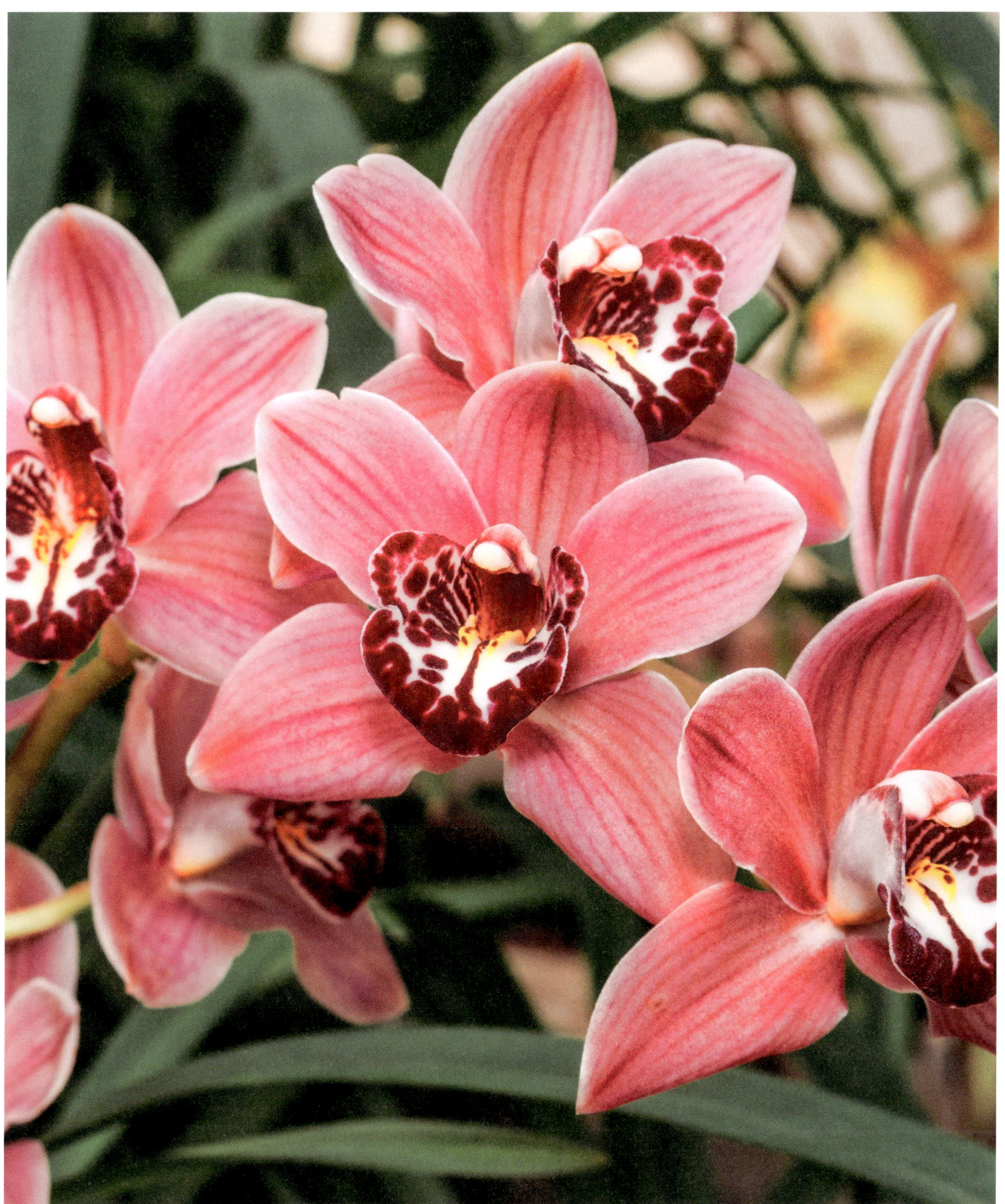

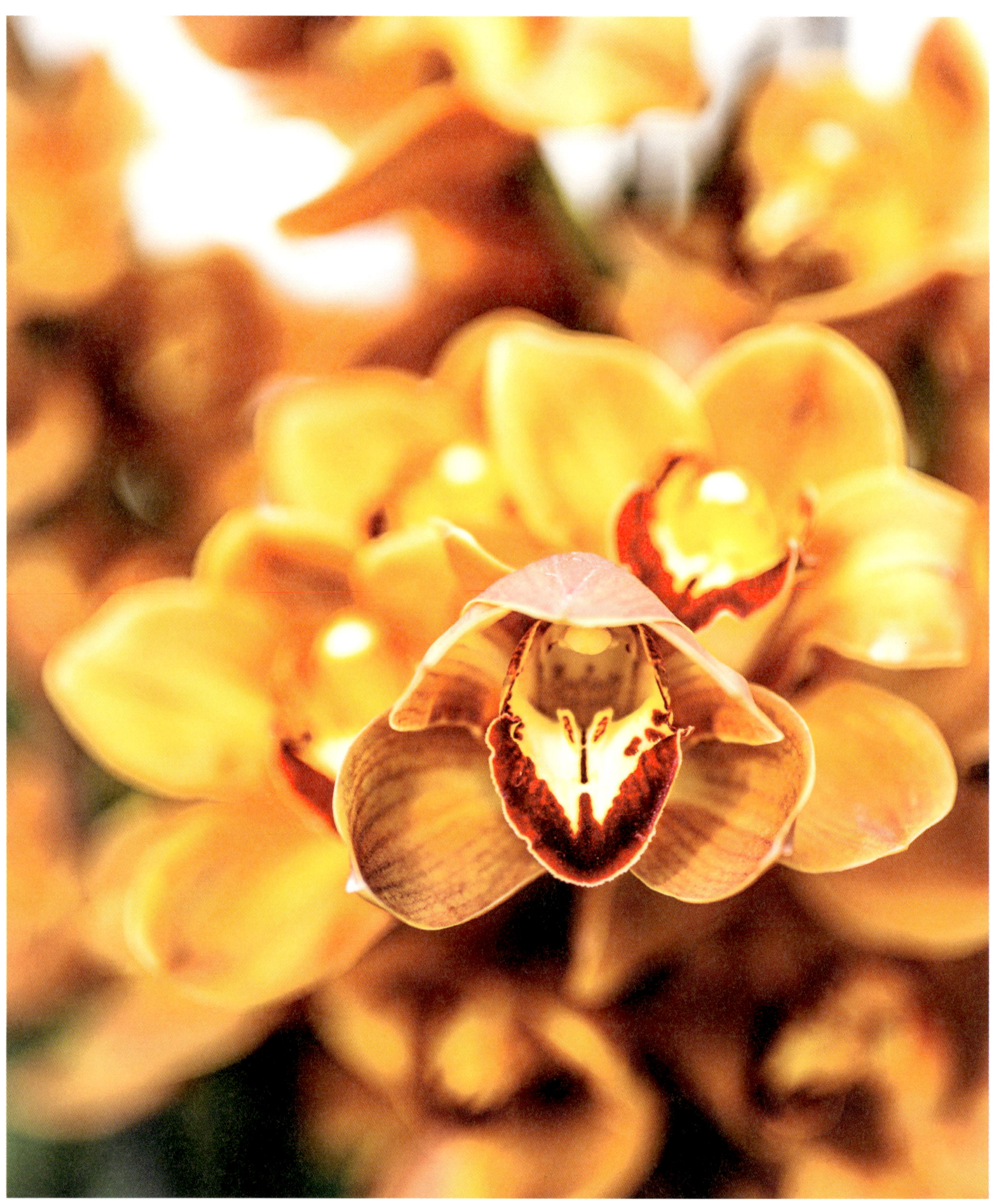

Platanthera or Phalaenopsis

A PROUD SPLASH OF ORANGE

Orange orchids are not common, which makes them a special sight. The vibrant color of their flowers symbolizes pride and success. So, for example, if you give an orange-flowering *Platanthera* or a *Phalaenopsis* as a gift in your professional life, you can join it with a well-deserved pat on the back and say, "Well done, keep up the good work!"

STOLZES ORANGE

In Orange erblühende Orchideen sind gar nicht so weit verbreitet und deshalb eine Besonderheit: Ihre leuchtende Farbe symbolisiert Stolz und Erfolg. Wer also z. B. im Berufsleben eine orangefarbene *Platanthera* oder eine *Phalaenopsis* verschenkt, kann damit auch ein anerkennendes Schulterklopfen ausdrücken: gut gemacht – und weiter so!

Oncidium or Phalaenopsis

BOLD YELLOW

In Asia, yellow orchids like *Oncidium* or *Phalaenopsis* are a symbol of rebirth. They convey optimism and hope for a fresh new beginning. However, yellow can send mixed messages in different cultures, so giving yellow orchids as a gift should be done with caution. While the color yellow is considered regal in China, yellow flowers may be understood as an expression of jealousy in France, and in some South and Central American cultures yellow flowers are placed on graves.

KRAFTVOLLES GELB

Gelbe Orchideen wie die *Oncidium* oder die *Phalaenopsis* gelten in Asien als Symbol der Wiedergeburt, vermitteln Optimismus und drücken die Hoffnung auf einen guten Neuanfang aus. Da Gelb in verschiedenen Kulturen allerdings widersprüchliche Farbsignale setzt, ist auch das Verschenken solcher Orchideen mit Vorsicht zu genießen – denn während in China Gelb als königliche Farbe gilt, könnten gelbe Blüten in Frankreich eher als Ausdruck von Eifersucht verstanden werden, und in manchen süd- und mittelamerikanischen Kulturen werden gelbe Blumen auf Gräber gelegt.

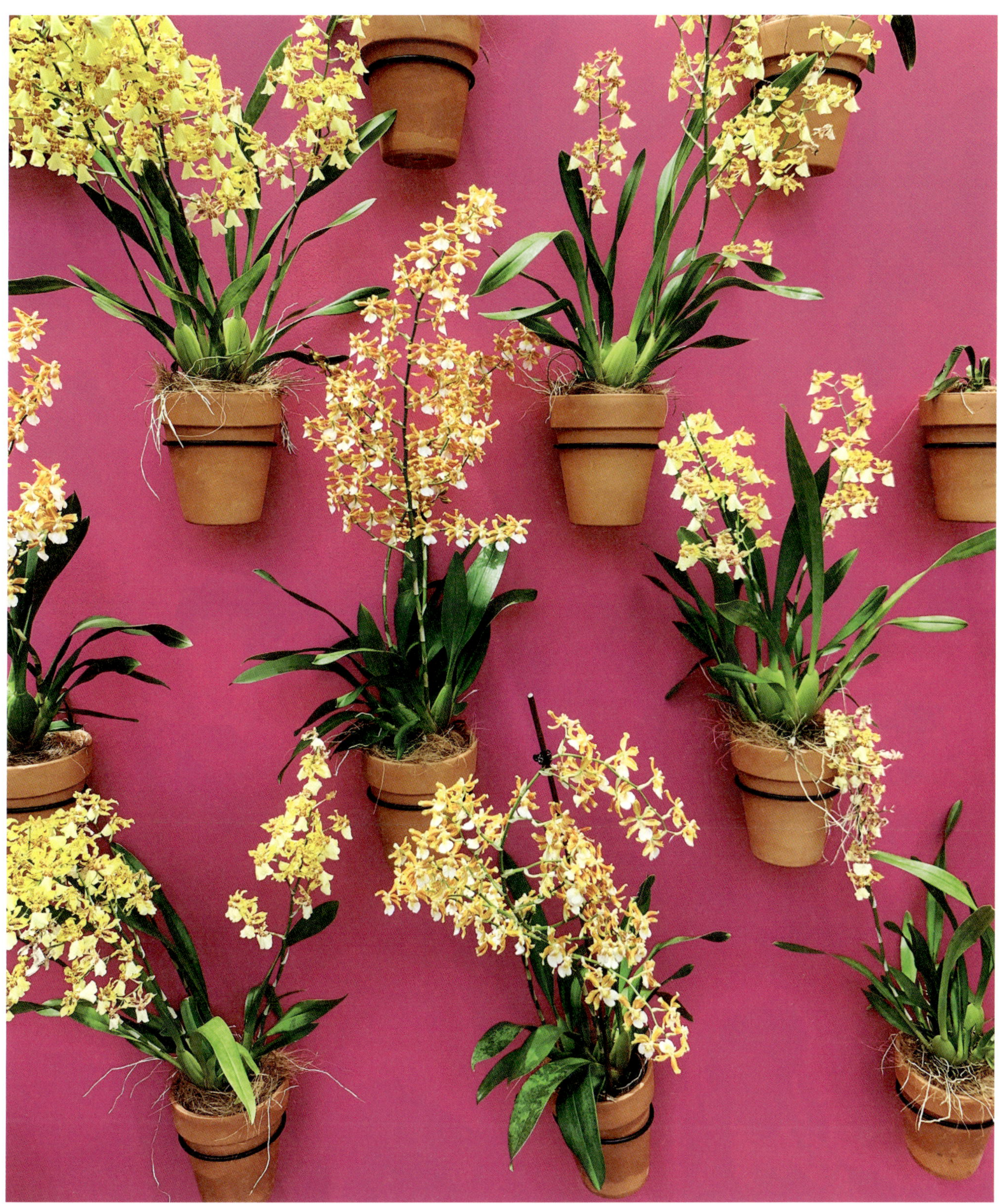

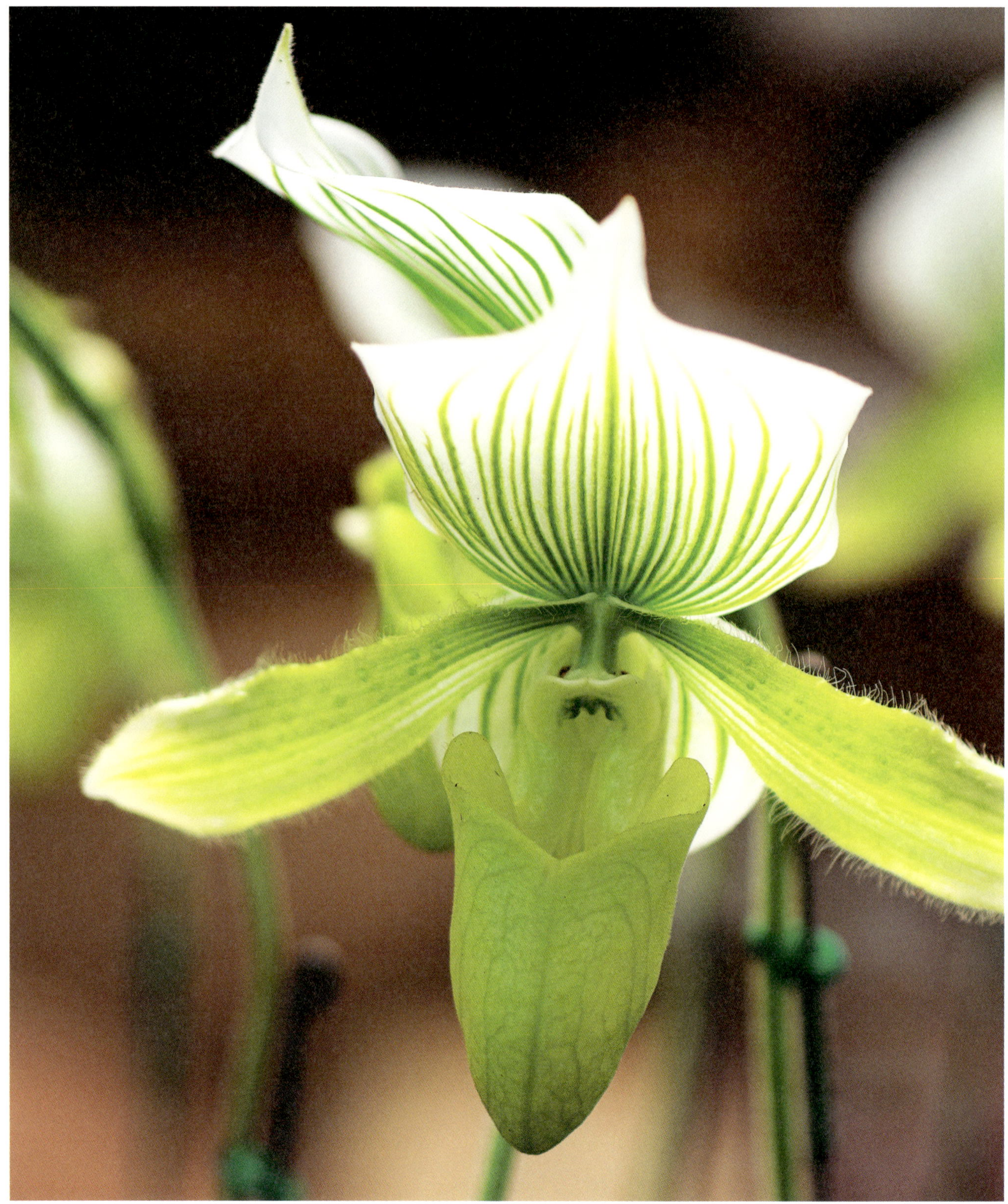

Paphiopedilum

HOPEFUL GREEN

Green is often associated with happiness, health, and prosperity, and the greenish hue of the Venus slipper orchid brings a sophisticated elegance that is sure to delight any flower enthusiast with a taste for the unusual. As green is also widely regarded as the color of hope in many cultures, a *Paphiopedilum* with green and white flowers is a thoughtful way to brighten the day of a loved one going through difficult times.

HOFFNUNGSFROHES GRÜN

Grün gilt als Farbe von Glück, Gesundheit und Wohlstand, und das Blütengrün eines Venusschuhs bringt eine stilvolle Eleganz mit sich, die jeder Blumenfreund mit einem Faible fürs Besondere zu schätzen weiß. Da Grün in vielen Kulturen auch als Farbe der Hoffnung gesehen wird, ist beispielsweise eine grün-weiß blühende *Paphiopedilum* auch eine schöne Geste, um damit liebe Menschen in schwierigen Zeiten aufzumuntern.

Phalaenopsis

INNOCENT WHITE

The *Phalaenopsis* blooms in the white of innocence and purity, making it a perfect gift for an occasion such as a christening. Its elegant white flowers are also ideal for a bridal bouquet. As one of the most popular orchid species in the world, the moth orchid is frequently chosen as a houseplant, thanks to its persistent blooms. It is an excellent component for clean, minimalist interior design concepts.

UNSCHULDIGES WEISS

Im Weiß der Unschuld und Reinheit blüht die *Phalaenopsis*: Bei feierlichen Anlässen wie einer Taufe ist die vielleicht beliebteste Orchideenart weltweit ein willkommenes Geschenk. Ihre weißen Blüten wirken aber auch in einem Brautstrauß wunderbar elegant. Dank ihrer ausdauernden Blütenpracht zählt die Schmetterlingsorchidee zu den willkommensten Zimmerpflanzen und passt ideal zu klaren, minimalistischen Einrichtungskonzepten.

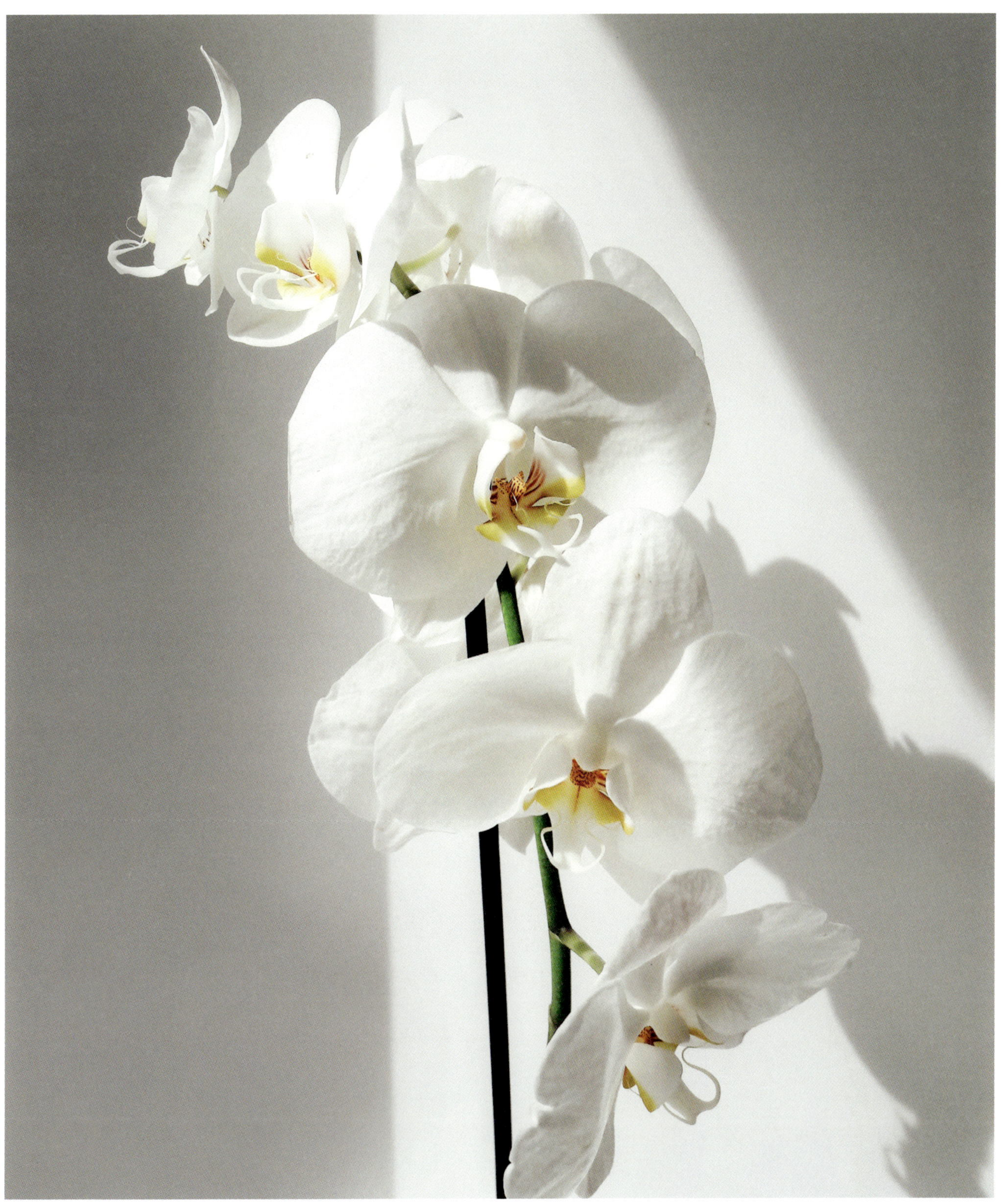

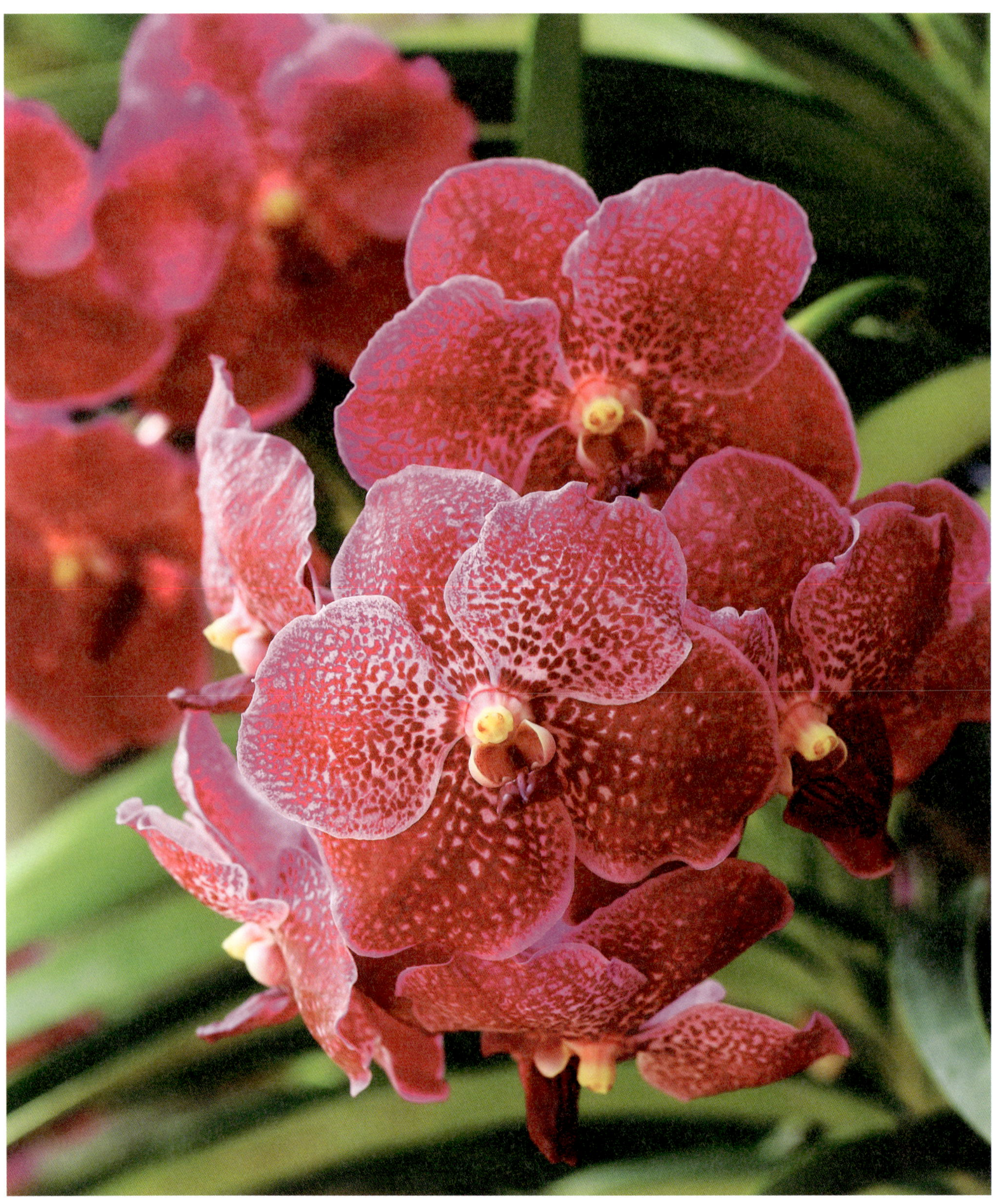

Vanda

CHEERFUL PINK

The grace and beauty of a pink *Vanda* orchid is simply indescribable. This vibrant flower makes for an ideal gift to congratulate new parents, bringing with it a life-affirming sense of cheerfulness. A young girl, on the other hand, may receive delicate light pink *Vanda* orchids, which could serve as a declaration of love from a bashful admirer.

HEITERES PINK

Die Anmut einer Orchideenblüte in Rosa ist unbeschreiblich: Als Geschenk für frischgebackene Eltern bringt beispielsweise das leuchtende Pink einer *Vanda* lebensbejahende Fröhlichkeit mit sich. Und wenn ein junges Mädchen zarte Orchideenblüten in hellem Rosa überreicht bekommt, mag dahinter womöglich eine Liebeserklärung eines schüchternen Verehrers stecken …

Dendrobium

REGAL PURPLE

The regal purple of the *Dendrobium* orchid is a symbol of respect and admiration. Giving purple orchids as a gift is a perfect way to show your appreciation. The luxurious shades of purple and violet have always been associated with nobility and religious dignitaries, making *Dendrobium nobile* an ideal choice to pay your respects or express your admiration to someone deserving.

NOBLES VIOLETT

Violette Orchideenblüten wie z. B. die der Traubenorchidee gelten als Zeichen der Ehrerbietung. Möchte man Wertschätzung vermitteln, sind purpurfarbene Orchideen das perfekte Geschenk. Da die Farbnuancen Violett und Purpur stets ein Hauch von Luxus umweht und sie seit langem mit Königen und kirchlichen Würdenträgern in Verbindung gebracht werden, ist eine *Dendrobium nobile* auf jeden Fall ein geeignetes Geschenk, um jemandem Respekt zu zollen oder Bewunderung auszudrücken.

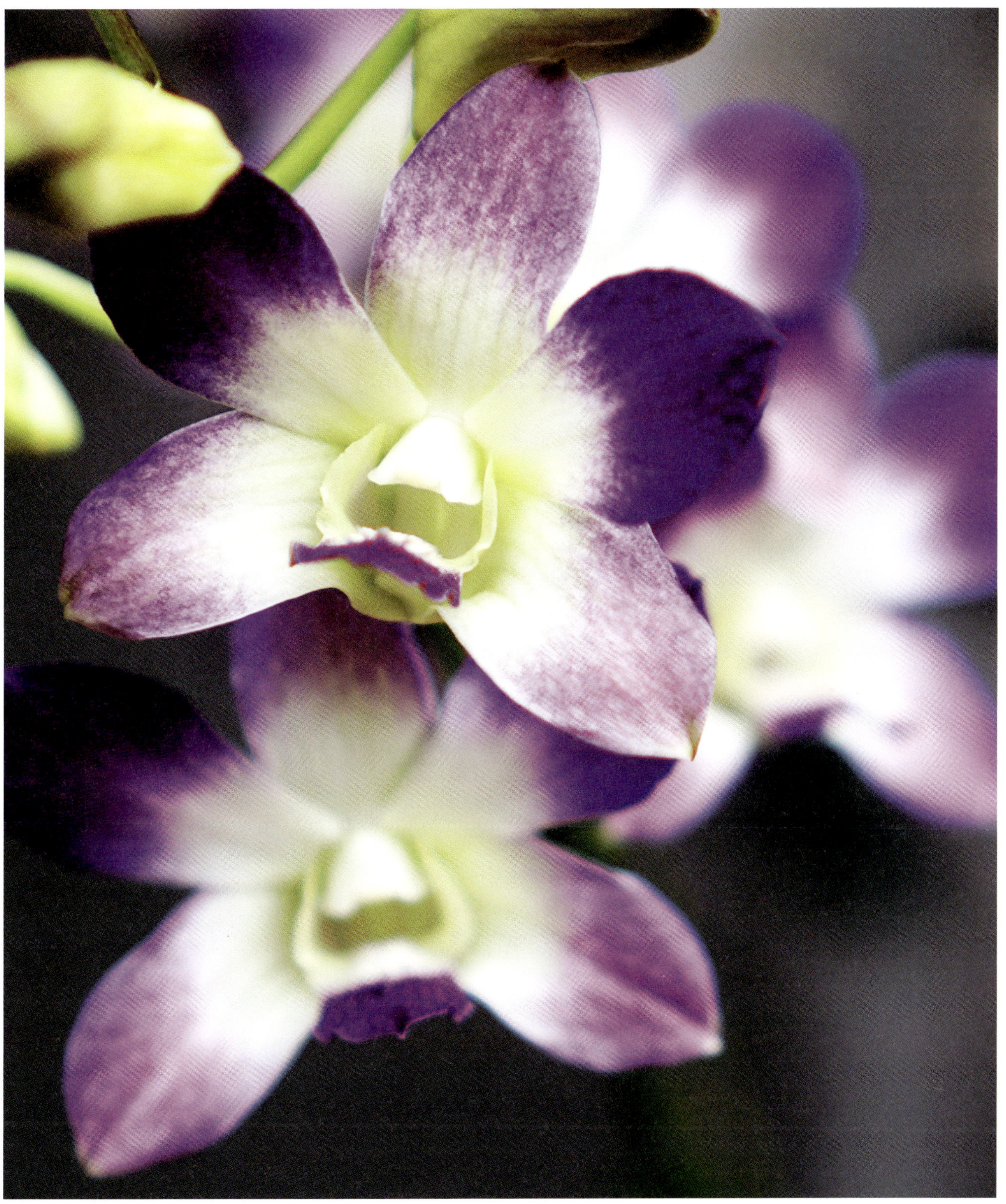

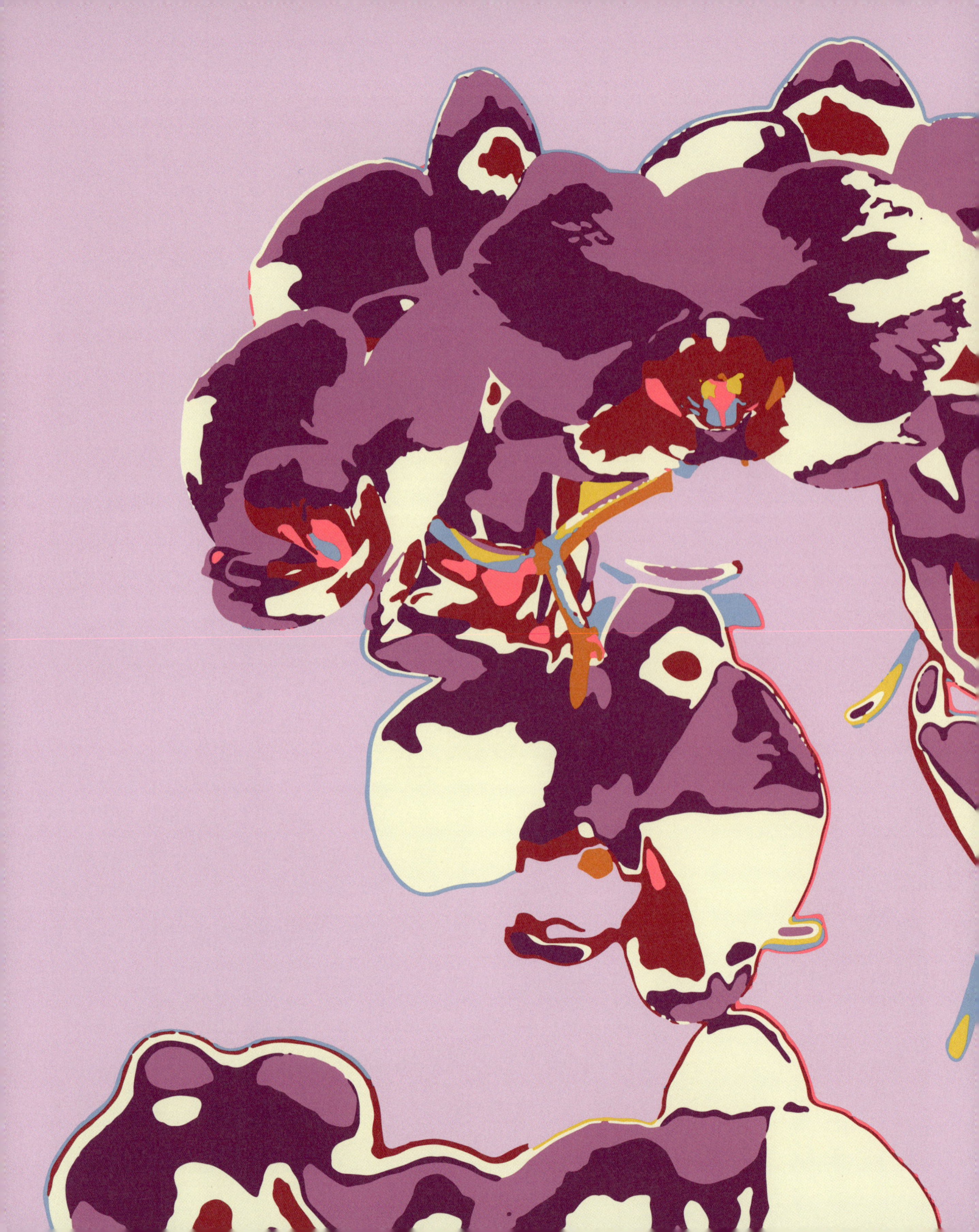

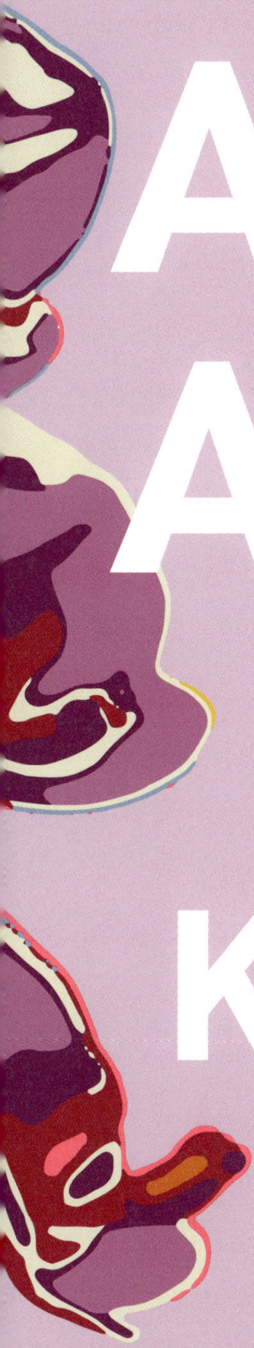

ART AND ARTISTS

KUNST UND KÜNSTLER

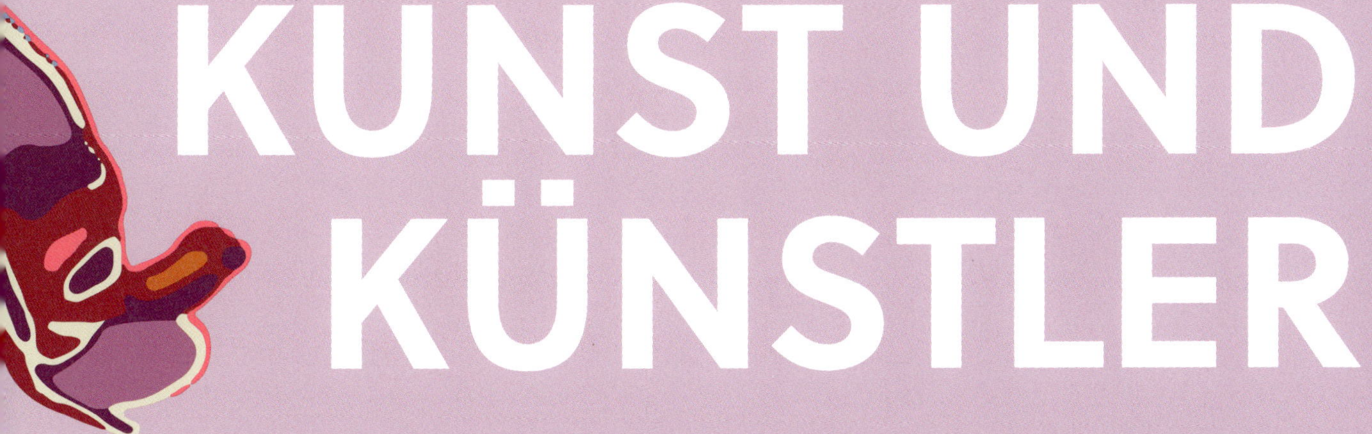

Heinrich Gustav Reichenbach and Henry George Moon

THE REICHENBACHIA

POSSIBLY THE MOST ELITIST-EVER ADVERTISEMENT FOR ORCHIDS

Heinrich Gustav Reichenbach, a German botanist, rose to prominence as Europe's leading orchid expert following the passing of his close friend and fellow scientist, John Lindley *(see page 166)*. The German-born orchid magnate, Henry Frederick Sander, who resided in England, published a now-famous, four-volume work on orchids, the *Reichenbachia*, in Reichenbach's honor. This impressive volume included life-size illustrations of nearly 200 orchid species. Despite its immense production costs, the Reichenbachia may be considered not only a standard botanical work but also perhaps the most extravagant of sales catalogues for Sander's thriving orchid trade. In a testament to Sander's exclusive clientele, the four volumes were dedicated to Queen Victoria of England, Empress Augusta Victoria of Germany, Tsarina Maria Feodorovna of Russia, and Queen Henriette of Belgium.

The colorful prints in the *Reichenbachia* were based on precise watercolors of orchids painted by Sander's son-in-law, Henry George Moon (1857–1905), who was perhaps more dedicated to the scientific ethic than his sales-oriented father-in-law. The watercolors that Moon painted for the *Reichenbachia* thus represent not idealized visions of their floral subjects but are absolutely lifelike depictions of plants he had meticulously studied.

FOLLOWING PAGES | Some of Henry George Moon's
fascinating drawings from the *Reichenbachia*

Heinrich Gustav Reichenbach und Henry George Moon

DIE REICHENBACHIA

DIE VIELLEICHT EXKLUSIVSTE ART, MIT DER ORCHIDEEN JEMALS BEWORBEN WURDEN

Der deutsche Botaniker Heinrich Gustav Reichenbach avancierte nach dem Tod seines Wissenschaftskollegen und engen Freundes John Lindley *(siehe Seite167)* zum führenden Orchideenexperten Europas. Ihm zu Ehren veröffentlichte der deutschstämmige, in England lebende Orchideenmagnat Henry Frederick Sander das berühmte vierbändige Orchideenwerk *Reichenbachia*, das durch seine lebensgroßen Darstellungen von fast 200 Orchideenarten beeindruckte. Man darf sich, trotz der immensen Produktionskosten, die *Reichenbachia* nicht nur als botanisches Standardwerk, sondern auch als sehr extravaganten Verkaufskatalog für Sanders florierenden Orchideenhandel vorstellen: Die Widmungen der vier Bände – an Englands Königin Victoria, die deutsche Kaiserin Augusta Victoria, die russische Zarin Maria Feodorovna und Königin Henriette von Belgien – zeigen, wie exklusiv Sanders Kundenkreis war.

Die farbenprächtigen Drucke der *Reichenbachia* wurden nach präzisen Orchideenaquarellen von Sanders Schwiegersohn Henry George Moon (1857–1905) erstellt. Dieser hatte sich dem Wissenschaftsethos wohl stärker verschrieben als seinem verkaufsorientierten Schwiegervater, und so sind Moons Aquarelle für die *Reichenbachia* keine Überzeichnung der Blütenschönheiten, sondern absolut naturgetreue Darstellungen der von ihm akribisch studierten Pflanzen.

NÄCHSTE SEITEN | Beispiele der wundervollen Zeichnungen von Henry George Moon für die *Reichenbachia*

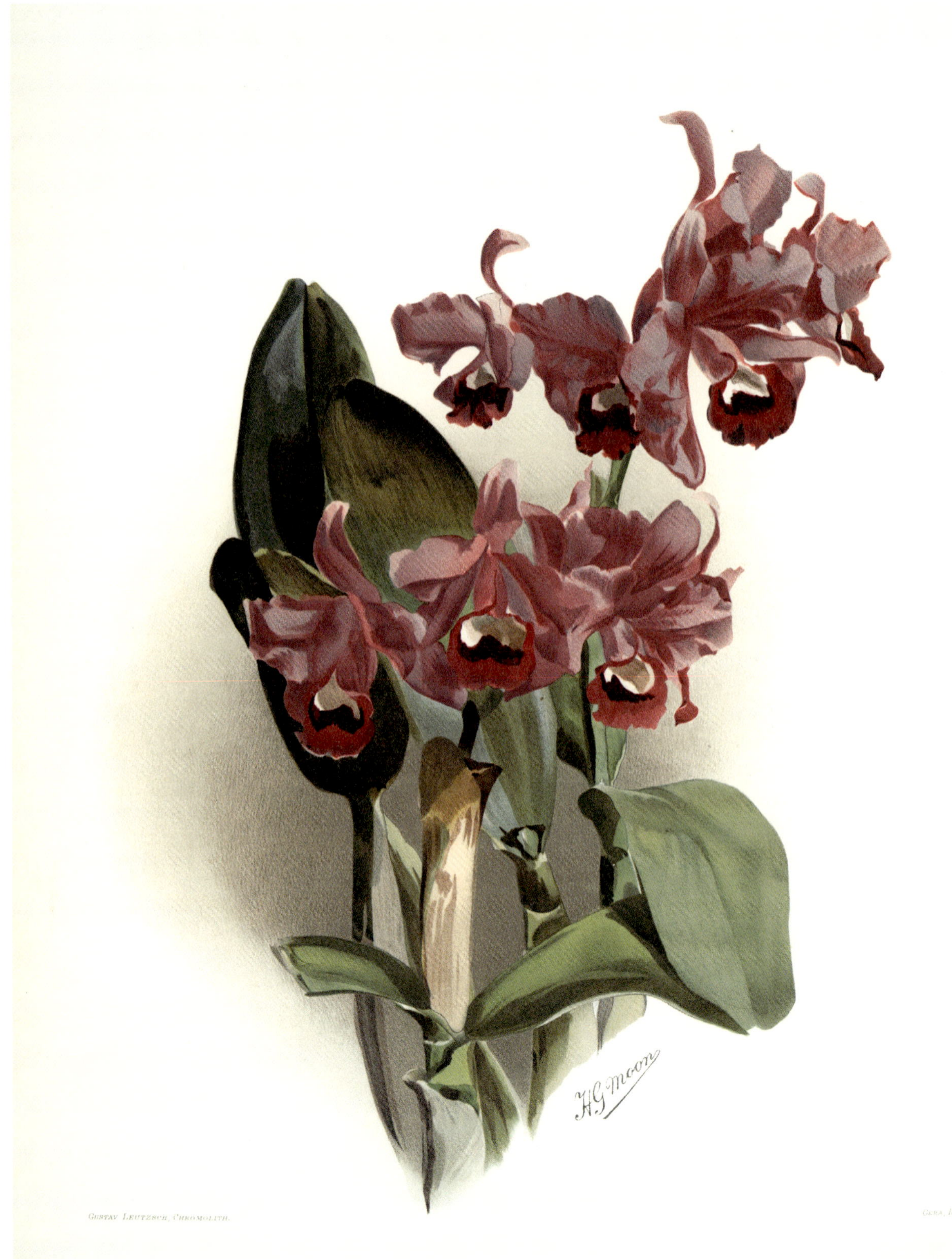

CATTLEYA BOWRINGIANA.

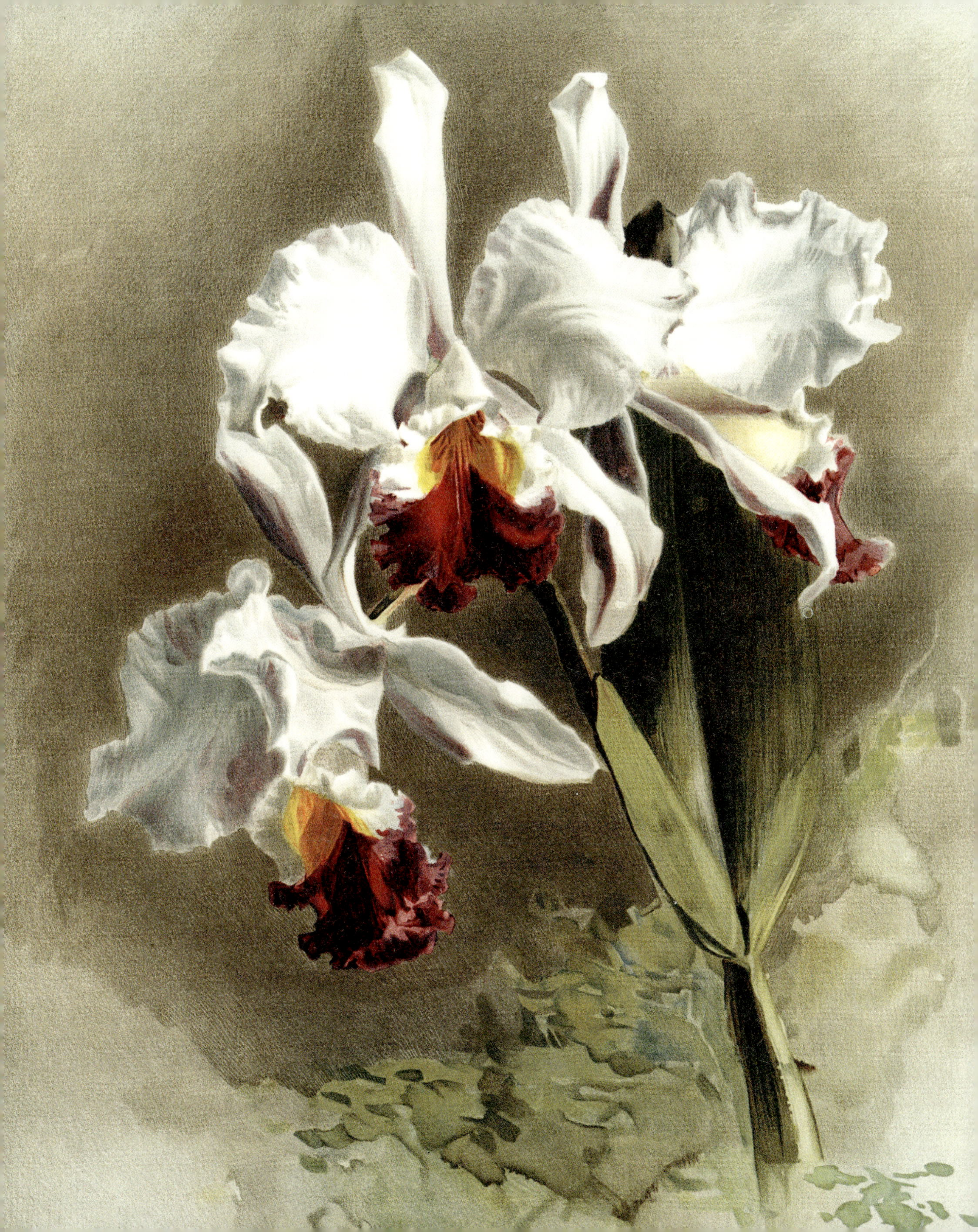

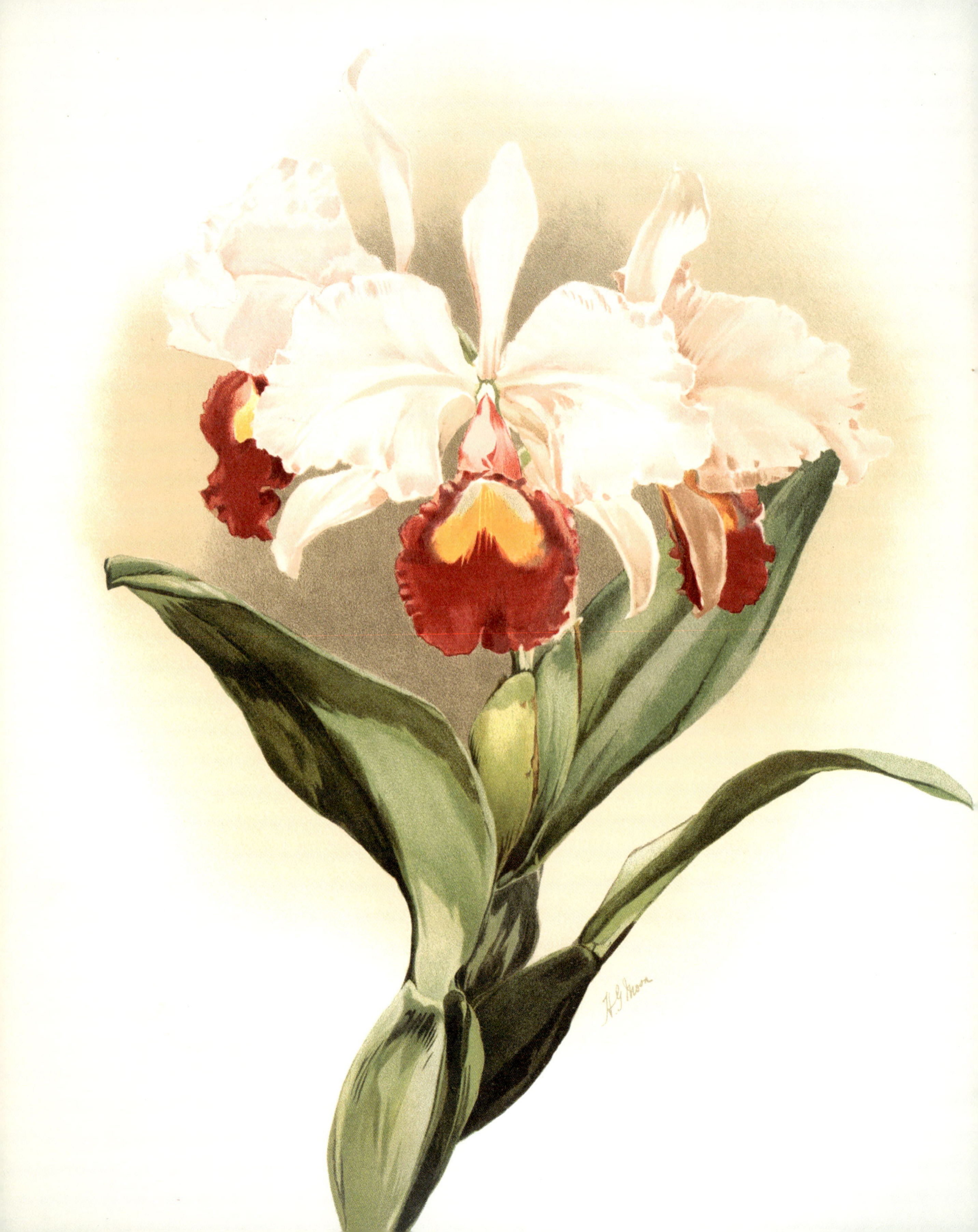

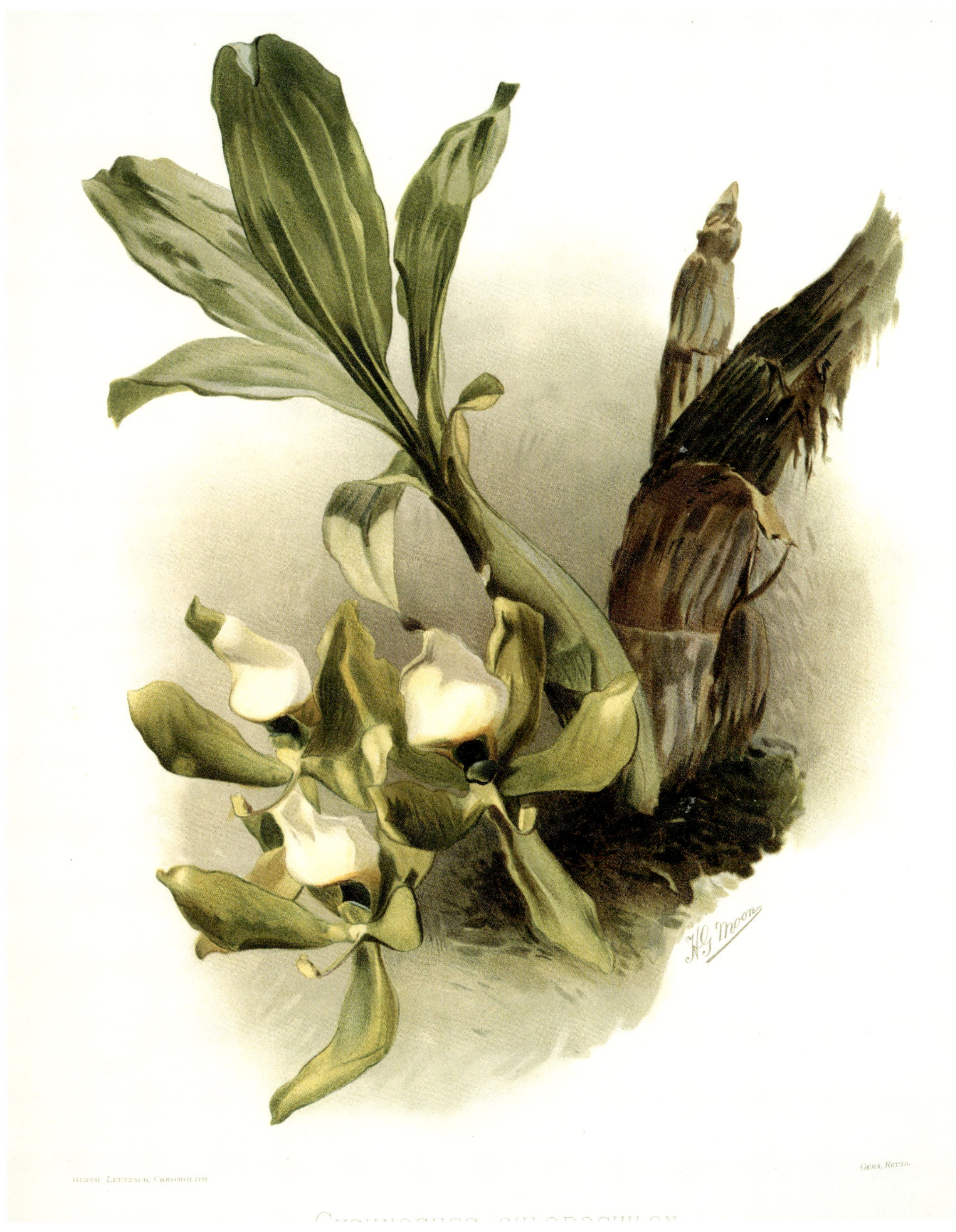

CYCNOCHES CHLOROCHILON

Hard work

SARAH-ANN DRAKE

THE ARTIST SARAH-ANN DRAKE REMAINS AS ELUSIVE AS HER WATERCOLORS OF ORCHIDS WERE PRECISE

It is both unfortunate and unusual that an artist as talented as Sarah-Ann Drake (1803–1857) should be so overshadowed by her own work. Little is known about her life though she was probably the finest orchid painter of her day. She came to the attention of orchid expert John Lindley through her childhood friendship with his sister, and it is believed that she joined the family as a governess to his children. However, Lindley soon discovered that Drake had an extraordinary artistic gift: her flower portraits are remarkable, not only for their realism and precision, but also for each image's graceful composition.

Lindley commissioned Drake to provide the illustrations for his monograph, *Sertum Orchidaceum (see page 166)*. Her exceptional artistry also won her commissions from a range of botanical magazines as well as from James Bateman's monumental *The Orchidaceae of Mexico and Guatemala (see page 162)*. She produced over 1,100 illustrations for a single publication alone, *The Botanical Register*. Drake retired when that journal was discontinued, and she spent the rest of her life in rural Norfolk, where she died at the early age of 53. In recognition of her remarkable talents, Lindley named a genus of hammer orchids, *Drakaea*, in her honor.

NEXT PAGE | By Sarah-Ann Drake:
Left *Odontoglossum grande*,
right *Sobralia macrantha*

Die Fleißige

SARAH-ANN DRAKE

SO PRÄZISE IHRE ORCHIDEENAQUARELLE SIND, SO SCHWER FASSBAR BLEIBT DIESE KÜNSTLERIN

Dass eine Künstlerin so sehr hinter ihrem Werk zurücktritt wie Sarah-Ann Drake (1803–1857), ist so ungewöhnlich wie bedauerlich: Nur wenig ist über das Leben der wohl talentiertesten Orchideenmalerin ihrer Zeit bekannt. Über ihre Jugendfreundschaft mit der Schwester des Orchideenspezialisten John Lindley kam sie in dessen Familie – vermutlich als eine Art Gouvernante für dessen Kinder. Dass Sarah-Ann Drake aber ein herausragendes künstlerisches Talent besaß, entdeckte der Orchideenforscher schon bald: Neben Realismus und Präzision zeichnet ihre Blumenporträts auch eine wunderbare Anmut in der Bildkomposition aus.

Von Lindley erhielt Drake den Auftrag, seine Monografie *Sertum Orchidaceum (siehe Seite 167)* zu illustrieren. Ihre außergewöhnlich kunstvolle Darstellung brachten der ungemein produktiven Künstlerin auch Aufträge für diverse Botanikmagazine und James Batemans Monumentalwerk *The Orchidaceae of Mexico and Guatemala (siehe Seite 163)* ein. Allein für *The Botanical Register* fertigte sie über 1100 Illustrationen an! Mit dessen Einstellung zog sich auch die Künstlerin zurück und verstarb mit nur 53 Jahren im ländlich abgeschiedenen Norfolk. Als Zeichen von Lindleys tiefer Wertschätzung benannte dieser nach Sarah-Ann Drakes Tod die Gattung der Hammerorchideen, *Drakaea*, nach der ihm ergebenen Künstlerin.

NÄCHSTE SEITE | Links eine *Odontoglossum grande*, rechts eine *Sobralia macrantha* von Sarah-Ann Drake

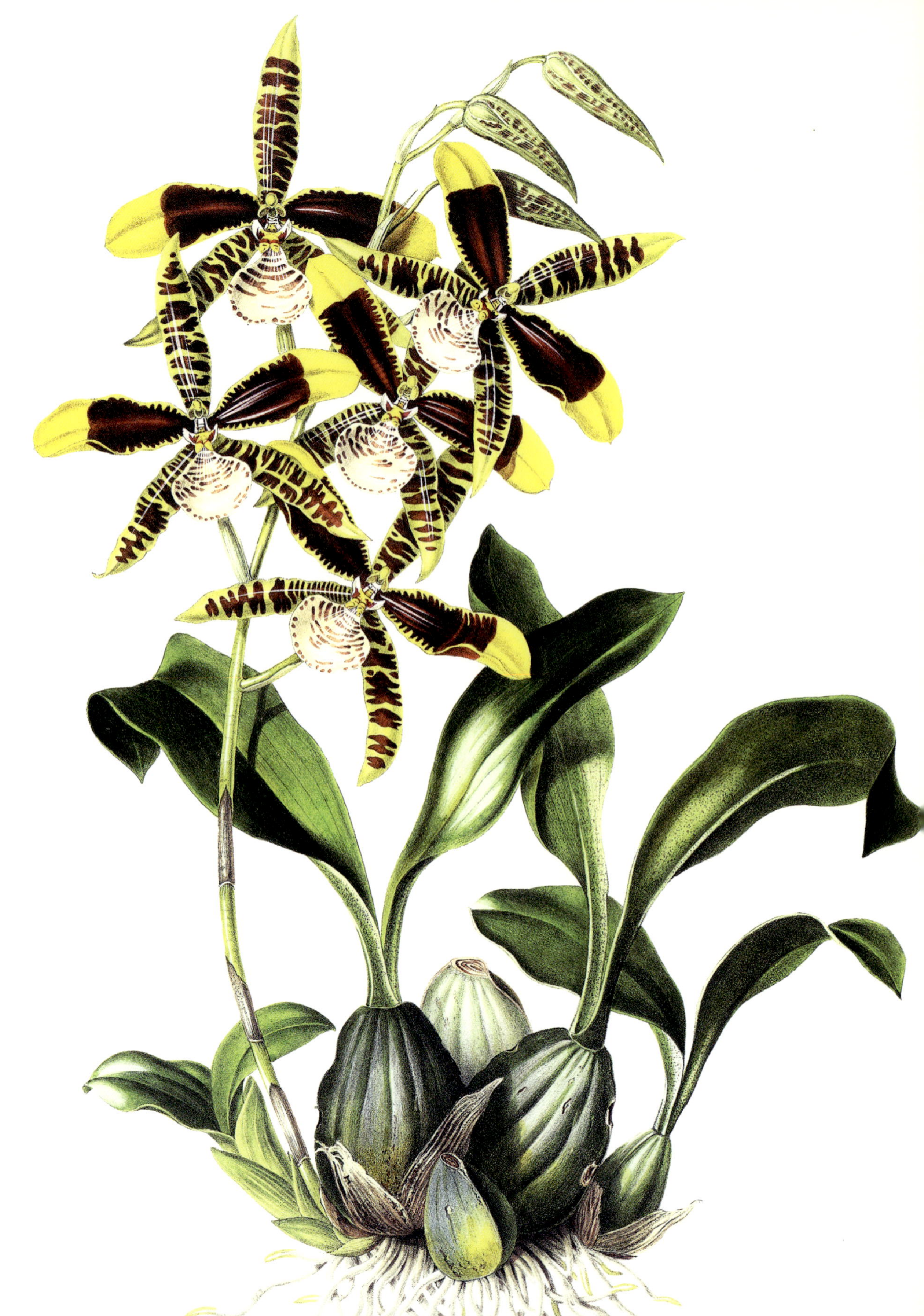

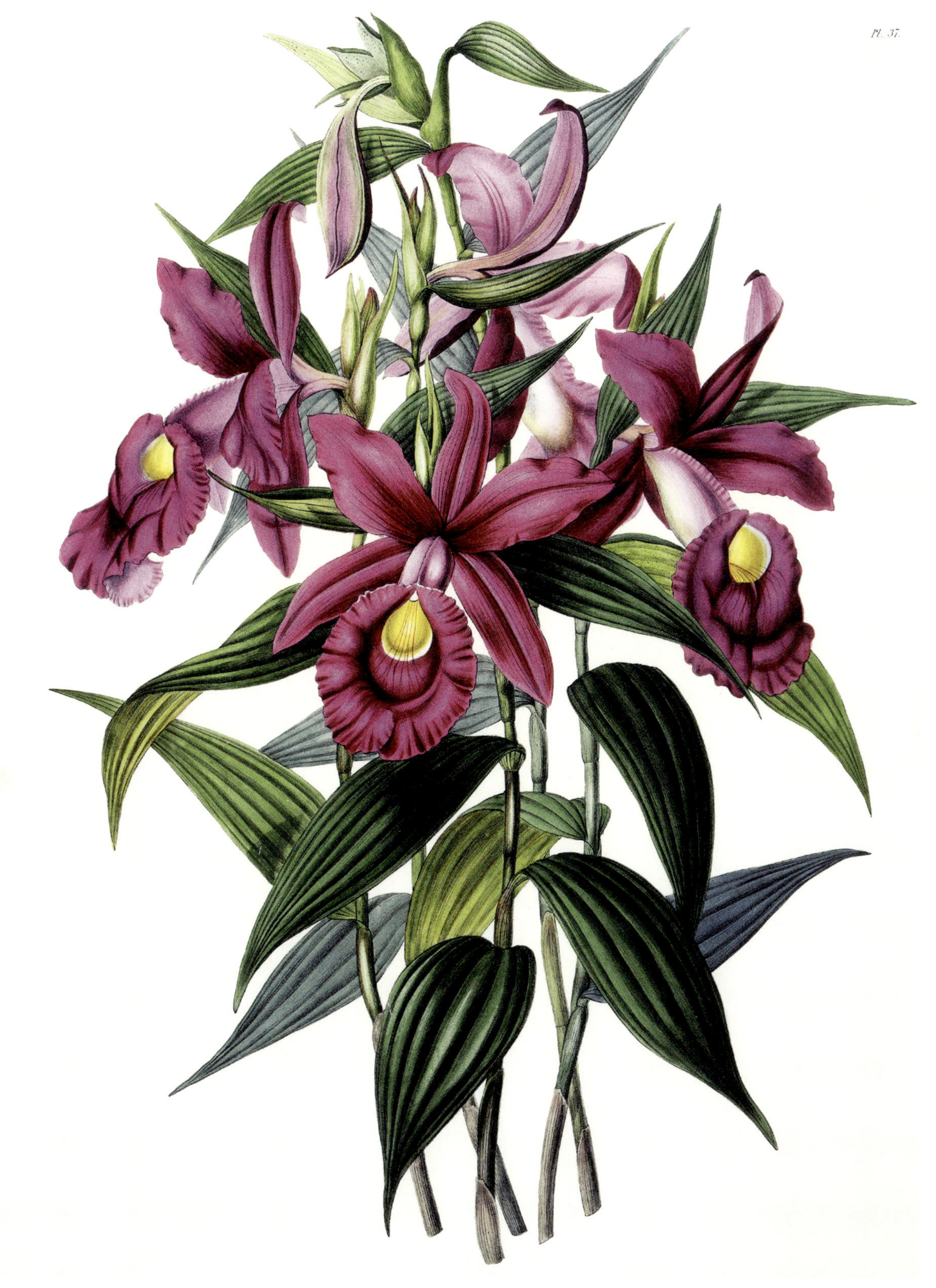

The adventurer

MARTIN JOHNSON HEADE

CATTLEYA ORCHID AND THREE BRAZILIAN HUMMINGBIRDS, 1871

Following in the footsteps of the orchid hunters, Martin Johnson Heade journeyed to South America to explore its exotic flora. With a keen eye for detail, he captured this extraordinary scene that appears to be both a snapshot of the landscape and an otherworldly still life. Amid the carefree buzzing-about of hummingbirds, a foreboding grey cloud gathers, blotting out the idyllic blue sky. The magnificent *Cattleya*, one senses, sits at the cusp of its final thunderstorm.

Der Abenteurer

MARTIN JOHNSON HEADE

CATTLEYA ORCHID AND THREE BRAZILIAN HUMMINGBIRDS, 1871

Auf den Spuren der Orchideenjäger reiste Martin Johnson Heade nach Südamerika, um dort die exotische Natur zu studieren. Sein Blick für Details ließ ihn diese außergewöhnliche Komposition schaffen, die halb lebendige Momentaufnahme und Landschaftsdarstellung, halb unwirkliches Stillleben zu sein scheint: Während die Kolibris noch selbstvergessen herumschwirren, verdrängt bedrohliches Grau den paradiesisch blauen Himmel – und man ahnt, dass die majestätische *Cattleya* ihrem letzten Gewittersturm bald nicht mehr standhalten kann.

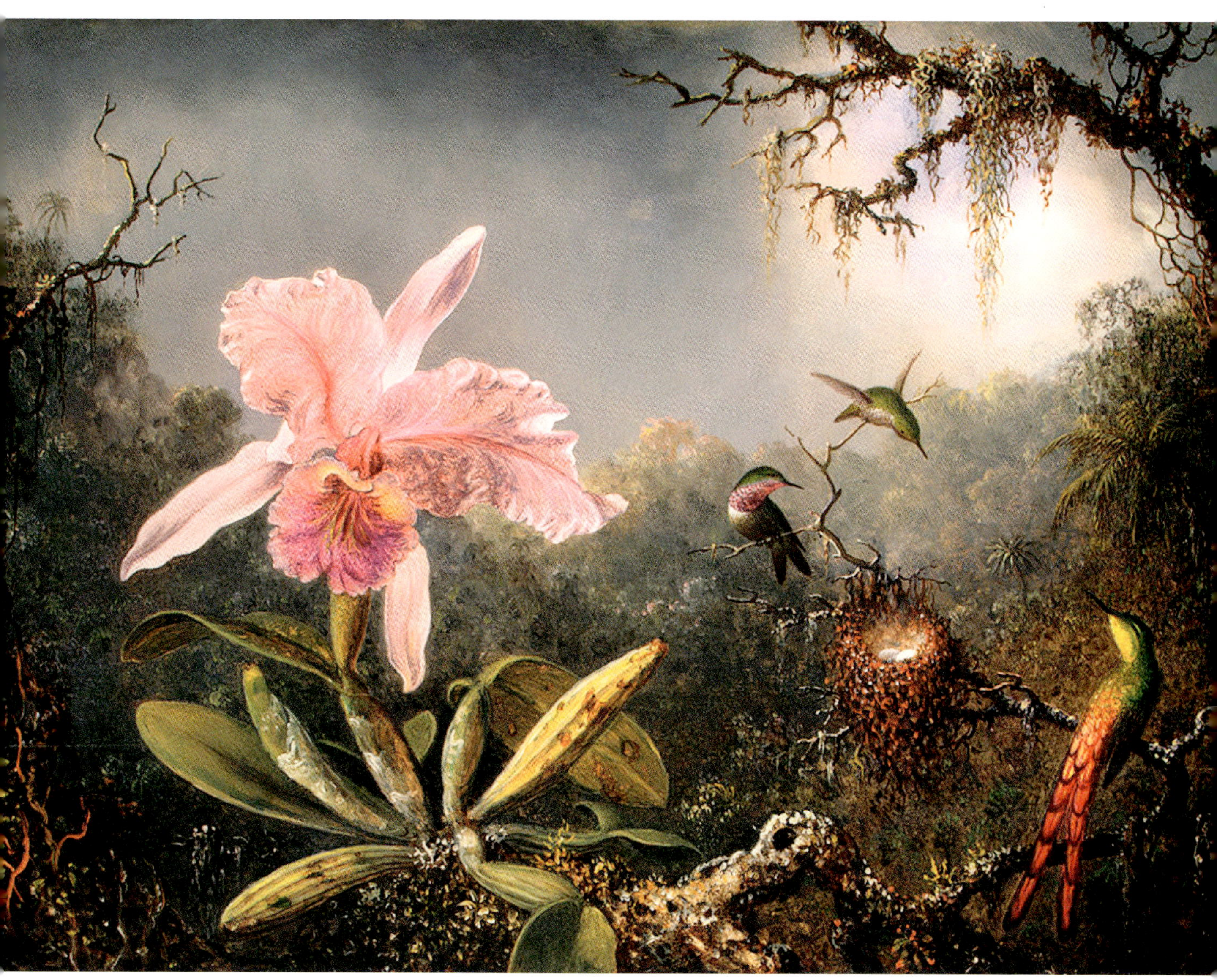

Congenial partners

ERNST HAECKEL & ADOLF GILTSCH

ORCHIDEAE IN *ART FORMS IN NATURE*

Today, the worldview and sociopolitical stance of zoologist Ernst Haeckel are obsolete. However, the watercolors in his renowned *Kunstformen der Natur (Art Forms in Nature)*, published between 1899 and 1904, speak for themselves. Lithographer Adolf Giltsch skillfully produced colored plates from them, merging art and science in their artistic compositions. Haeckel's depiction of "Venus flowers" – his term for *Orchideae* – significantly influenced the visual aesthetics of numerous Art Nouveau artists.

Kongeniale Partner

ERNST HAECKEL & ADOLF GILTSCH

ORCHIDEAE IN DEN *KUNSTFORMEN DER NATUR*

In Weltanschauung und sozialpolitischen Ansichten ist der Zoologe Ernst Haeckel heute umstritten, aber die Aquarelle zu seinen berühmten *Kunstformen der Natur* – erschienen zwischen 1899 und 1904 – stehen in ihrer Schönheit für sich. Der Lithograf Adolf Giltsch erstellte daraus meisterhaft farbenprächtige Bildtafeln, deren kunstvolle Arrangements Malerei und Wissenschaft verschmelzen ließen. So hatte auch die Darstellung der „Venusblumen", wie Haeckel die *Orchideae* nennt, prägenden Einfluss auf die Bildästhetik vieler nachfolgender Jugendstilkünstler.

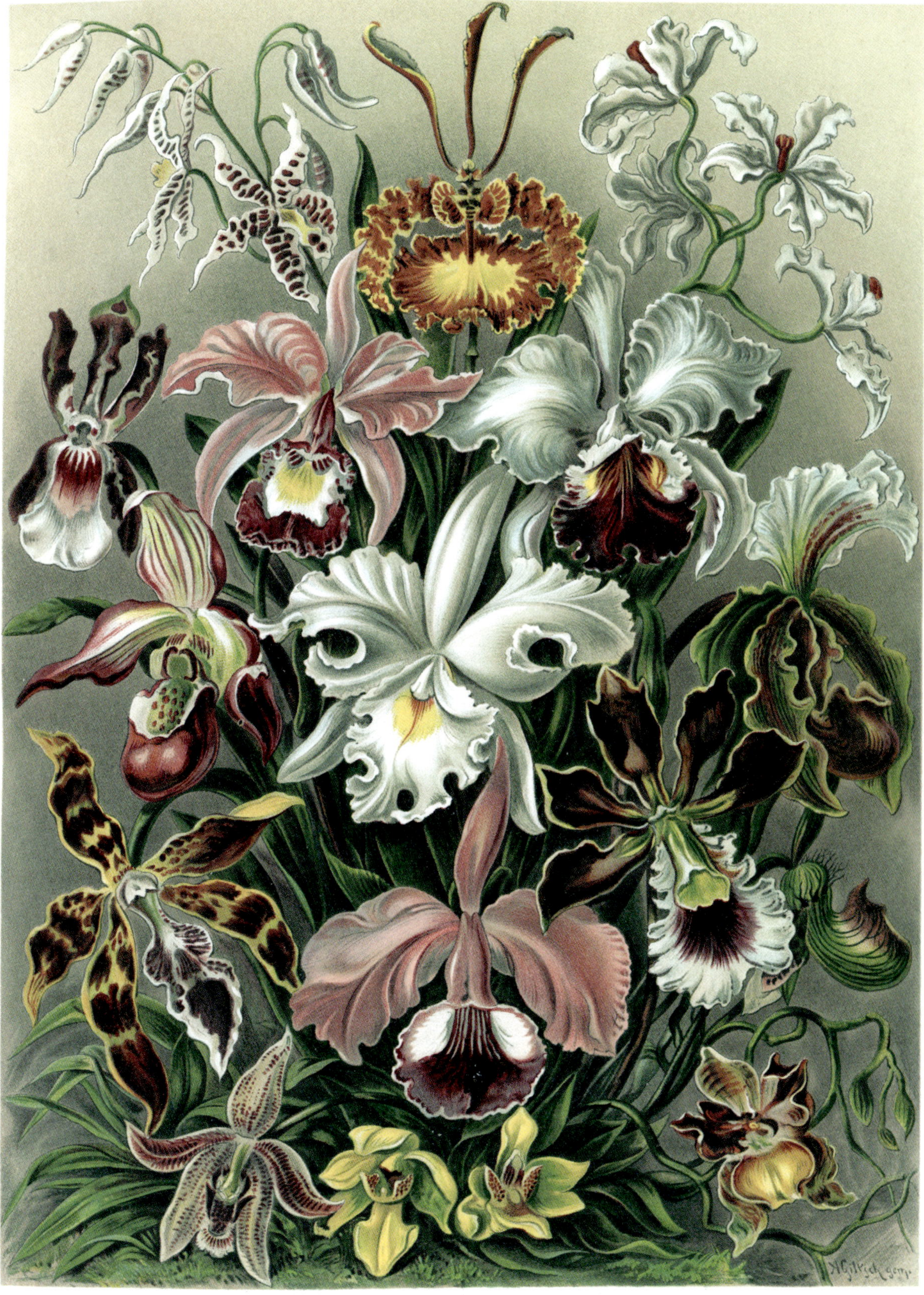

Gently needling your friend

GEORGIA O'KEEFFE

NARCISSA'S LAST ORCHID, 1940

Georgia O'Keeffe is well-known for her sensual and elegant portrayals of flowers. She often zooms in on the flower itself so closely that it dissolves into abstractions. One of her works, *Narcissa's Last Orchid*, features a white *Cattleya* given to her by her friend, Narcissa Swift King. The painting's title is a reference to King herself.

King, who had given O'Keeffe flowers, had expressed disappointment that her gift hadn't been duly appreciated, and she said it would be the last time she ever gave O'Keeffe an orchid. Despite this, the two women maintained a close friendship despite, or perhaps even because of, King's barbed dedication.

Liebevoller Seitenhieb an eine Freundin

GEORGIA O'KEEFFE

NARCISSA'S LAST ORCHID, 1940

Georgia O'Keeffe ist für ihre sinnlich-eleganten Blumendarstellungen berühmt: Sie zoomt so nahe an den Blütenstand heran, bis dieser sich in abstrakte Formen auflöst. *Narcissa's Last Orchid* zitiert O'Keeffes Freundin Narcissa Swift King, von der die abgebildete weiße *Cattleya* stammte: Da sie ihre Blumengabe nicht genügend gewürdigt sah, wäre dies nun die letzte Orchidee, die sie O'Keeffe jemals schenken würde! Die Freundschaft der beiden war trotz – oder gerade dank – der pointierten Widmung von Dauer.

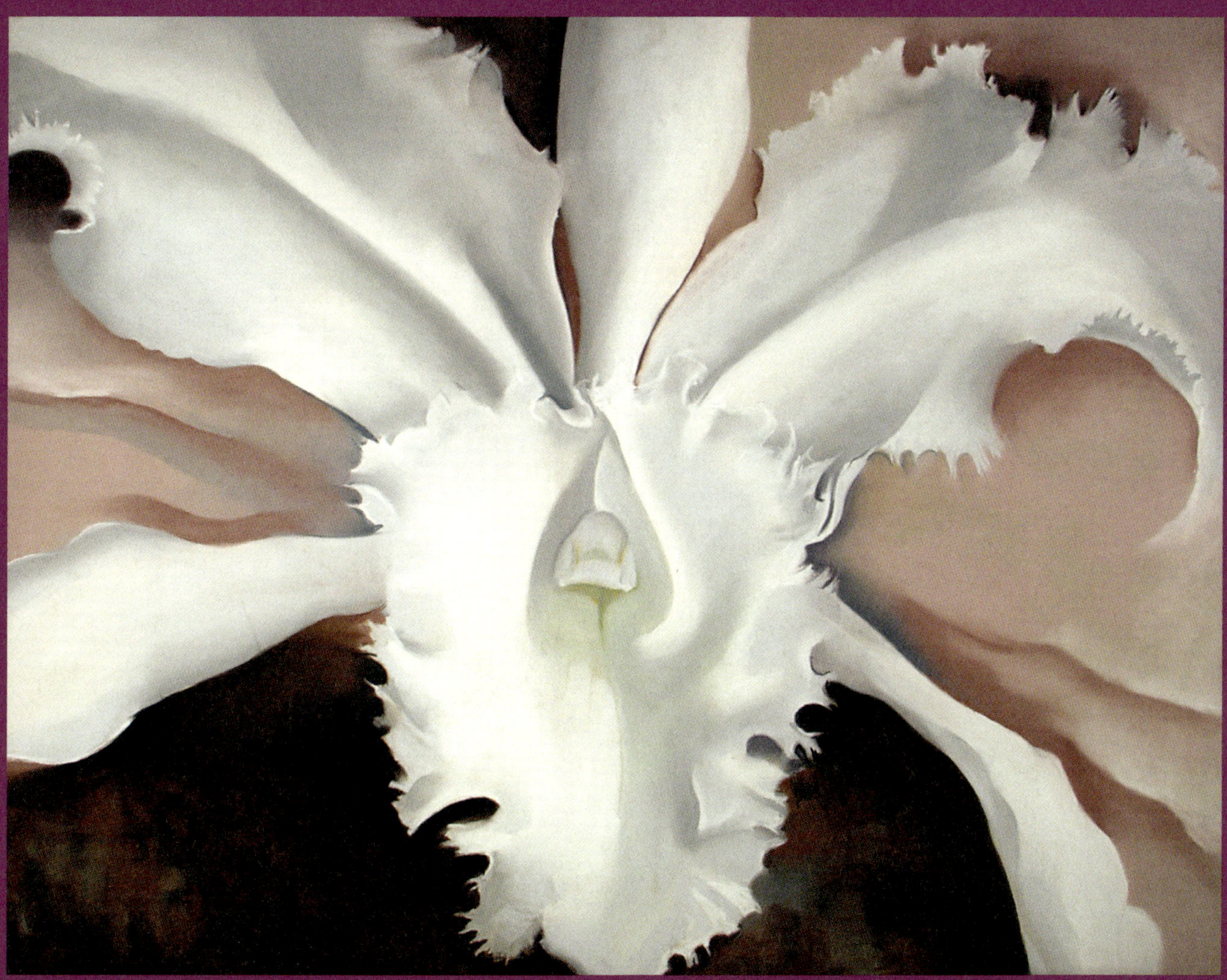

Anja Klaffenbach wurde die Liebe zur Natur mehr oder weniger in die Wiege gelegt: Im Garten ihrer Mutter, einer passionierten Blumenfreundin und Freizeitgärtnerin, entdeckte sie schon von klein auf die große Vielfalt der heimischen Pflanzenwelt. Nachdem ihr Studium der Anglistik und Romanistik sie nach England führte, war ihre Faszination für Gartenkunst besiegelt. Mittlerweile hat sie sich in ihrem Zuhause in einer bayerischen Großstadt den Traum vom eigenen Garten verwirklicht.

Wenn sie nicht gerade als freie Texterin und Autorin Bücher über Reisen und Pflanzenvielfalt schreibt, erprobt sie dort ihren grünen Daumen mit dem Erkunden neuer und altbekannter Gemüse- und Zierpflanzen.

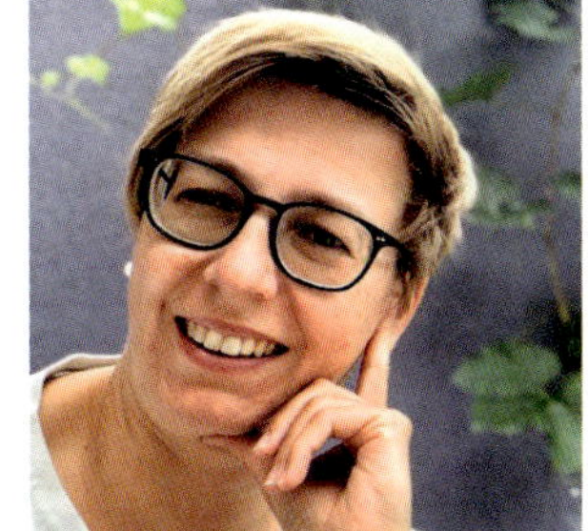

Anja Klaffenbach was more or less born with a love of nature: In the garden of her mother, a passionate flower enthusiast and amateur gardener, she discovered the great diversity of the local flora from an early age. After her studies of English and Romance languages and literature had brought her to England, her fascination for the art of gardening was sealed. In the meantime, she has realised her dream of having her own garden in her home in a large Bavarian city. When she is not busy as a freelance copywriter and author writing books about travelling and botanical life, she can be found there, putting her green thumb to the test by exploring new and old-fashioned species of edible and ornamental plants.

© 2023 teNeues Verlag GmbH

Text: Anja Klaffenbach
English Translation: John Augustus Foulks
Picture Editing: Heide Christiansen
Design & Copy Editing: Eva Stadler
Proofreading: Gesa Wendhack
Editorial Coordination: Dr. Johannes Abdullahi, teNeues Verlag
Production: Sandra Jansen, teNeues Verlag
Photo Editing, Color Separation: Jens Grundei, teNeues Verlag

ISBN German Cover: 978-3-96171-482-7
ISBN English Cover: 978-3-96171-548-0
Library of Congress Number: 2023941898
Printed by GPS in BiH

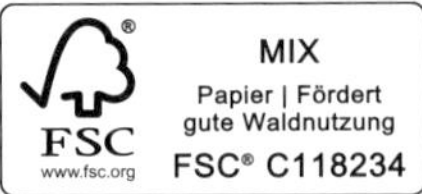

Bibliographic information published by the Deutsche Nationalbibliothek:
The Deutsche Nationalbibliothek lists this publication in the Deutsche Nationalbibliografie; detailed bibliographic data are available on the Internet at dnb.dnb.de.

Published by teNeues Publishing Group

teNeues Verlag GmbH
Ohmstraße 8a
86199 Augsburg, Germany

Düsseldorf Office
Waldenburger Straße 13
41564 Kaarst, Germany
e-mail: books@teneues.com

Augsburg/München Office
Ohmstraße 8a
86199 Augsburg, Germany
e-mail: books@teneues.com

Berlin Office
Lietzenburger Straße 53
10719 Berlin, Germany
e-mail: books@teneues.com

Press Department
e-mail: presse@teneues.com

teNeues Publishing Company
350 Seventh Avenue, Suite 301
New York, NY 10001, USA

www.teneues.com

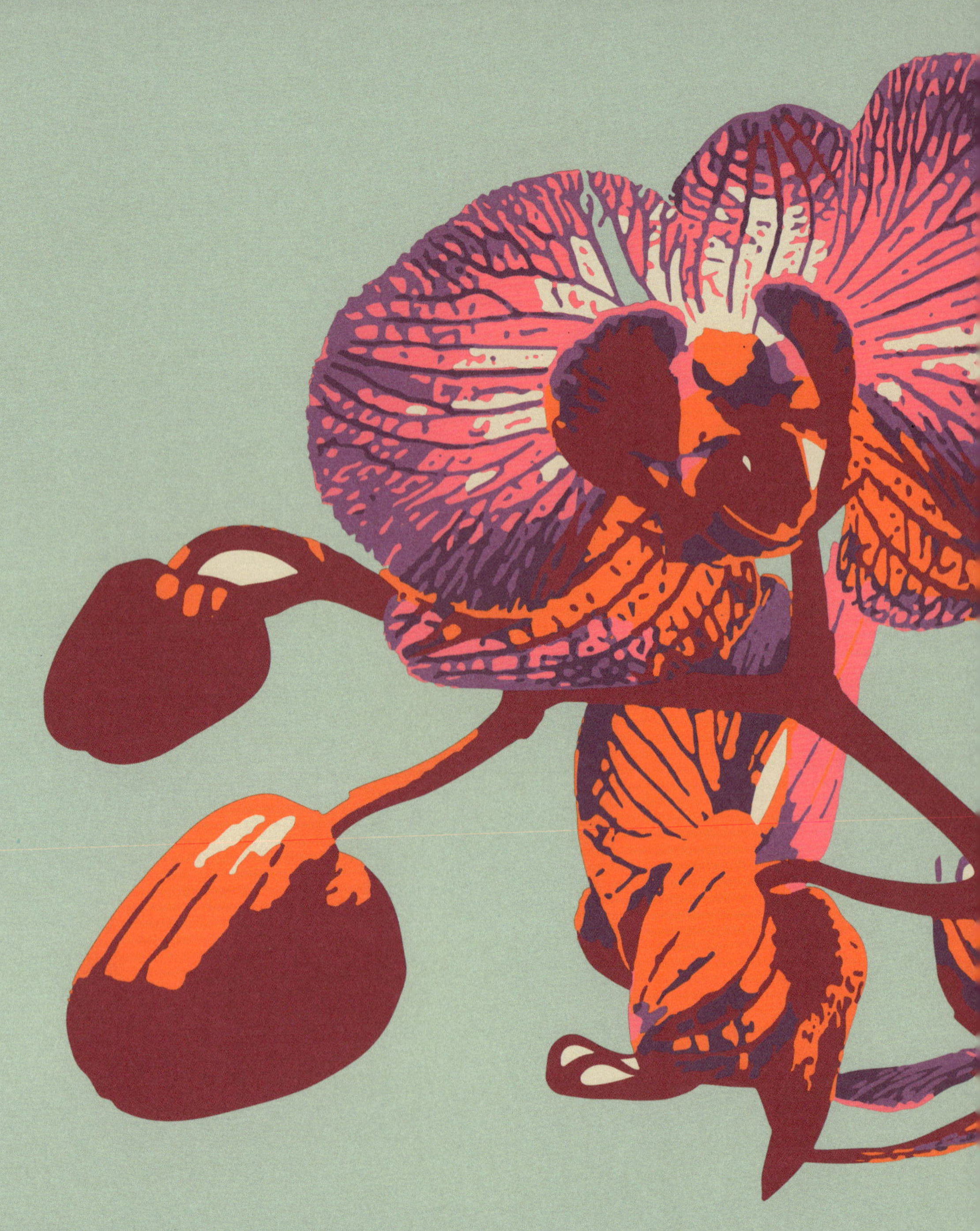